Springer **M**onographs in **M**athematics

More information about this series at http://www.springer.com/series/3733

Leonid Bunimovich • Benjamin Webb

Isospectral Transformations

A New Approach to Analyzing
Multidimensional Systems and Networks

Leonid Bunimovich
School of Mathematics
Georgia Institute of Technology
Atlanta, USA

Benjamin Webb
Department of Mathematics
Brigham Young University
Provo, UT, USA

ISSN 1439-7382　　　　ISSN 2196-9922 (electronic)
ISBN 978-1-4939-1374-9　　ISBN 978-1-4939-1375-6 (eBook)
DOI 10.1007/978-1-4939-1375-6
Springer New York Heidelberg Dordrecht London

Library of Congress Control Number: 2014944753

Mathematics Subject Classification: 05C82, 37N99, 65F30, 15A18, 34D20

© Springer Science+Business Media New York 2014
This work is subject to copyright. All rights are reserved by the Publisher, whether the whole or part of the material is concerned, specifically the rights of translation, reprinting, reuse of illustrations, recitation, broadcasting, reproduction on microfilms or in any other physical way, and transmission or information storage and retrieval, electronic adaptation, computer software, or by similar or dissimilar methodology now known or hereafter developed. Exempted from this legal reservation are brief excerpts in connection with reviews or scholarly analysis or material supplied specifically for the purpose of being entered and executed on a computer system, for exclusive use by the purchaser of the work. Duplication of this publication or parts thereof is permitted only under the provisions of the Copyright Law of the Publisher's location, in its current version, and permission for use must always be obtained from Springer. Permissions for use may be obtained through RightsLink at the Copyright Clearance Center. Violations are liable to prosecution under the respective Copyright Law.
The use of general descriptive names, registered names, trademarks, service marks, etc. in this publication does not imply, even in the absence of a specific statement, that such names are exempt from the relevant protective laws and regulations and therefore free for general use.
While the advice and information in this book are believed to be true and accurate at the date of publication, neither the authors nor the editors nor the publisher can accept any legal responsibility for any errors or omissions that may be made. The publisher makes no warranty, express or implied, with respect to the material contained herein.

Printed on acid-free paper

Springer is part of Springer Science+Business Media (www.springer.com)

To Larissa and Rebekah

Foreword

This book provides a new approach to the analysis of networks and, more generally, to those multidimensional dynamical systems that have an irregular structure of interactions. Here, the term *irregular structure* means that the system's variables depend on each other in dissimilar ways. For instance, all-to-all or nearest-neighbor interactions have a *regular* structure, because each variable depends on the others in a similar manner.

In practice, this structure of interactions is represented by a graph, called the network's *graph of interactions* or the network's *topology*. Depending on the particular network, this graph may be directed or undirected, weighted or unweighted, with or without loops, with or without parallel edges, etc. In each case, the techniques provided in this book can be directly applied to these networks.

It is worth mentioning that although these methods are fairly new, they have already proven to be an efficient tool in some classical and more recent problems in applied mathematics. Here, these techniques are presented as a way in which to view and analyze real-world networks.

One of the major goals of this book is to make these methods and techniques accessible to researchers who deal with such networks. With this goal in mind, we note that the computations required to implement these techniques are remarkably straightforward. In fact, they can be carried out using any existing software sophisticated enough to perform elementary linear algebra.

In terms of the book's content, we note that each of the results is given with a mathematical proof. However, with the hope that this book will be read and used as well by nonmathematicians, the text is written so that those interested in applications can safely ignore these arguments and use the stated formulas and techniques directly. Still, we stress the fact that only a basic understanding of linear algebra is needed to understand the proofs.

The definitions and results we present are motivated and accompanied by many examples, which both nonmathematicians and mathematicians should appreciate.

The book also contains a large number of examples and figures depicting the graphs associated with particular networks as well as their various transformations.

Almost all of these examples deal with directed graphs. The reason is that directed graphs are more general objects than undirected graphs. However, the theory developed here works just as well for undirected graphs. This is important, for instance, in the study of real networks, since a large number of those networks have an undirected graph structure (topology).

Because real-life networks are often large and have a complicated structure, it is tempting to find ways of simplifying them in terms of both their size and complexity. What is important, though, is that some basic or fundamental property of the network be preserved in this process. Yet such an attempt seems doomed to failure. These are real networks, so we do not know much if anything about them, including which characteristic(s) we should retain. Moreover, there are potentially many ways in which a network could be reduced. Hence, there is first the problem of choosing which way the network should be reduced and second determining what the reduced network tells us. Thus many objections are immediately raised if one wants to reduce the size of a network.

From this point of view, our goal of reducing a network may seem overly ambitious. In fact, one could ask how it is possible even to represent an arbitrary network. The universally excepted answer is that this can be done by drawing a graph whose vertices (nodes) correspond to the network elements and whose edges (links) correspond to the directed interactions between these elements.

Equivalently, one can represent a network by a matrix A with entries A_{ij}. In this representation, A_{ij} is the *strength* or *weight* of the directed interaction between the ith and jth network elements, where $A_{ij} = 0$ if these elements do not interact. Such a matrix is called the *weighted adjacency matrix* of a network. If the network's interaction strengths are not known, the nonzero entries of the matrix are set equal to 1, and A is called the (unweighted) *adjacency matrix* of the network. In practice, knowledge of a network's adjacency matrix is often the most one can hope to have.

It is well known that a very basic characteristic of a matrix is its spectrum, i.e., its collection of eigenvalues including multiplicities. One of the main questions we address in this book is whether it is possible to reduce a network to some smaller network while preserving the network's spectral properties. Phrased another way, this question could be stated as whether it is possible to reduce the size of a network's adjacency matrix while maintaining the network's eigenvalues.

The immediate answer to this question is, of course, no. In fact, while presenting these results, we have had audience members protest that what we hope to do is impossible. Indeed, as everyone knows, the fundamental theorem of algebra states that an $n \times n$ matrix has n eigenvalues, while a smaller matrix has fewer.

However, our claim is that it is possible to reduce a matrix and preserve the matrix's spectral properties. In this book, we show that the answer to our question becomes yes if one considers a larger class of matrices, namely matrices with entries that are rational functions of a spectral parameter λ. That is, it is possible to reduce an $n \times n$ matrix with scalar entries to a smaller $m \times m$ matrix with functions as entries and maintain the matrix's spectrum. We refer to this process as isospectral matrix reduction.

At this point, the reader may think that by isospectrally reducing a network's adjacency matrix we are, in fact, shifting the complexity of a network's structure (topology) to the complexity of its edge weights. We pause to reassure our readers that we have considered this idea and that many facts and results in this book demonstrate that such is not the case. However, before moving on, we stress just one fundamental fact regarding isospectral reductions.

The structure (topology) of an isospectrally reduced network does not depend on the strengths (weights) of the initial unreduced network. It depends only on the network's structure. The structure of the reduced network will be the same regardless of the strengths of interactions in the initial network. Therefore, the isospectral reductions we consider really capture some hidden but intrinsic information regarding the structure of a network.

This approach to analyzing networks is based on ideas and methods from the theory of dynamical networks, which is a part of the modern theory of dynamical systems. The first dynamical networks addressed in this theory were the so-called coupled map lattices (CML). CMLs were introduced in the mid-1970s, almost simultaneously, by four physicists in four countries. The mathematical theory of CML was begun in [7], in which the first precise definitions of space-time chaos and a coherent structure were given. Nowadays, the theory of lattice dynamical systems is a respected part of contemporary dynamical systems theory (see, e.g., [13]).

A number of remarkable findings of the late 1990s demonstrated that real networks have very complicated topologies [3, 19, 20, 23, 24, 30]. The first thought was that the ideas of dynamical systems theory and of statistical mechanics could be applied to such systems, as had been done in [7, 12] for CMLs. However, infinite lattices have a group translation property, which is missing if a graph of interactions has an irregular structure. An approach to dealing with this irregular structure was eventually developed in [1, 4], in which the following was observed.

Every dynamical network has three features: (i) the individual dynamics of the network elements, e.g., a single isolated neuron in a neural network; (ii) the interactions between the elements of a network; and (iii) the structure (or topology) of a network. In this framework, we assume that a network's structure does not change over time, so that it has a fixed structure of interactions. However, as we later point out, the transformations considered in this book could be useful for studying networks that do have a structure that evolves over time.

Observe that features (i) and (ii) of a network are dynamical systems. Thus, it is customary to deal with such systems by analyzing the combined influence of (i) and (ii), as is done in other spatially extended systems. Perhaps the most popular example of these is reaction–diffusion systems in which nonlinear reactions push the system towards chaotic behavior while diffusion has a stabilizing effect. However, the question is what to do with (iii), which is clearly a static characteristic of the network.

As is shown in [1], the topology of a network can also be treated as a dynamical system generated by considering all infinite paths on the network's graph of interactions. This, together with the ideas from the theory of spatially extended systems, forms the basis of our approach.

This approach, as demonstrated in this book, has allowed for a variety of new results regarding dynamical systems, matrices, polynomials, and in particular, the stability of dynamical networks [8–10]. Most importantly, these techniques provide a reliable yet flexible tool for the analysis of real networks or conversely, the design of networks with specific properties. We also demonstrate, via the proofs we give, that this approach is internally consistent, which makes this an especially promising tool for analyzing real networks.

In this regard, consider, for instance, a network with the set elements (vertices) V. If the network is reduced to a smaller network with elements $A \subset V$, then this reduced network can generally be reduced to a network with an even smaller set of elements $B \subset A \subset V$. However, one could imagine that the initial network could be reduced over another set of elements $C \subset V$ and that the resulting network could be reduced over the same final set of elements $B \subset C \subset V$.

The question is, would these two isospectral reductions of the initial network onto a network with the set of elements B be identical? This, of course, is a serious consideration if our approach is to be useful. If the two results are different, then the utility of our approach would be questionable, to say the least.

In fact, one of our results demonstrates that these two isospectral reductions are the same. Moreover, the graph that results from any sequence of reductions depends only on the last (minimal) set of elements over which the network is reduced. This is another indication that our approach is internally consistent and that it identifies some intrinsic characteristic of a network's topology.

Additionally, it suggests at least two immediate possibilities regarding the analysis of real networks. First, it is possible to reduce a network isospectrally over any collection of elements of the initial network. The major question then is, who should choose this subset of elements, i.e., which elements of a network are the most important? Naturally, the expert in the field, e.g., biologist, medical doctor, engineer, is the most logical candidate for this task. An expert can both choose the elements over which to reduce a network and interpret the meaning of the reduced network.

Another potential use of these techniques is the following. There exist roughly 10 to 15 characteristics of a network's elements (vertices) and interactions (edges) that are routinely used in the analysis of real networks. These include centrality, in- and out-degree, betweenness, etc. Our theory allows anyone to determine the *core* of a network, i.e., a collection of elements that are the most important from the point of view of any such characteristic. We show that for any characteristic that uniquely describes a subset of elements (or edges), there is a unique reduced network with these and only these elements (or edges).

Such core subnetworks can, for instance, be complete graphs or graphs with nearest-neighbor connections, that is, a graph in which all vertices (or edges) are similar with respect to a particular characteristic. One can then get an expert's interpretation of the meaning of this core and compare it with other cores obtained using different characteristics.

Another general finding is that a rule that uniquely selects a collection of network elements (or edges) defines a partition of the set of all networks into classes of spectrally equivalent networks. The idea is that spectrally equivalent networks have similar dynamics. This equivalence means that if the networks L and M and the networks M and N are spectrally equivalent with respect to a given rule, then the networks L and N are also spectrally equivalent with respect to that rule.

It is important to mention here that we require such rules to define a subset of elements (or interactions) uniquely. For instance, the rule "remove one element of the network" does not uniquely specify which element should be removed. However, the rule "remove all elements (vertices) with minimal centrality" uniquely defines the set of elements to be removed. Therefore, this second rule partitions the set of all networks into spectrally equivalent classes, while the first does not.

It is worth mentioning that an isospectral reduction with respect to a network's interaction (edges) is absolutely analogous to isospectral reduction with respect to its elements (vertices). One need only consider the dual of the initial network's graph of interactions in which all vertices become edges and edges become vertices.

Having mentioned some of the applications of our approach, we would like to explain here what is principally new in our approach. In this respect, we have already mentioned that this approach requires the use of matrices with entries that are rational functions. But there is also something new from the point of view of graph theory, which is the theory so often used in the modern theory of networks.

We introduce the idea that there are special subsets of a graph's (network's) vertices (elements) over which the graph can be reduced. We call these special subsets the graph's *structural sets*. A collection of vertices is a structural set if its complement does not contain any cycles of the graph apart from *loops*, which are cycles of a single vertex. Every graph has at least one structural set. Moreover, in the case of an undirected graph, such sets have a specific form as a result of the graph's symmetries.

For different applications, we will sometimes use modifications or simplifications of this notion, but structural sets are at the heart of our theory. Once a structural set has been chosen, the entire graph (network) can be decomposed into a number of branches. These *branches* are paths between two vertices of the structural set that do not contain any other vertices of that set. Our procedure of isospectral reduction is, in a nutshell, the removal of all vertices that do not belong to the structural set. This is done by substituting each branch by a single edge and calculating the weights of these new edges.

It is important to note that in this procedure, we do not simply erase all vertices that are not in a structural set. We also add some new edges that were not present in the initial (larger) graph. Therefore, an isospectrally reduced graph is not a subgraph of the initial graph but a graph with a smaller number of vertices. Again, the same operation can be applied to edges of our graph instead of its vertices if we wish to reduce the graph isospectrally to a graph with fewer edges.

Having described, to some extent, what is new in our approach, we now give a detailed description of the book's content. The first chapter deals with isospectral matrix reductions. Formally, one of the major goals of the book is to develop a set of tools that will allow one to compress a network in a way that preserves all information relative to its spectrum. In this regard, we are technically dealing with an isospectral reduction of a network's (weighted) adjacency matrix. Thus, in Chap. 1 we describe how a matrix can be reduced.

Chapter 2 is the most fundamental chapter of the book. It describes the procedure of isospectral network reduction first introduced in [9]. The fundamental concepts of a structural set, branches, and branch weights are introduced there. This chapter also develops the operations of branch expansion, branch merging, and branch reweighting and demonstrates how these operations are useful for the analysis of dynamical networks.

In particular, the operations of expanding, reweighting, and merging a network's branches allow one to transform a network isospectrally while keeping the network's edge weights restricted to some set. Mathematically, such sets must be unital subrings of the ring of rational functions. For example, the weights 0 and 1 form such a ring, as do the positive real numbers. That is, it is not necessary for isospectral transformations, including those that reduce the size of a network, to result in a network that has edges with weights as fancy as rational functions.

Chapter 3 deals with the global stability of nondelayed and time-delayed dynamical networks. By *global stability* we mean the existence of a globally attracting fixed point in a network's phase space. In this particular context, we are interested not in the entire spectrum of a dynamical network but rather in its spectral radius, i.e., the maximum absolute value of all eigenvalues. Therefore, we introduce other transformations that are simpler than the isospectral transforms of Chap. 2. These transformations either preserve the spectral radius of a network or modify it in a specific way [10].

The first of these transformations that preserves the network's spectral radius is called an isoradial transformation. The second is referred to as a bounded radial transformation, which allows one to transform the network in such a way that its spectral radius remains below a certain value. By making use of these transformations and some structural features of networks, it is possible to obtain stronger sufficient conditions on a dynamical network's global stability than those obtained in [1]. These improved estimates come from taking into account the local structure of a network's topology, particularly its branch structure.

One intrinsic feature of a real network is that its dynamics are time-delayed. Such delays are typically caused by the "physical" distance between network elements as well as by the time required for a network element to process incoming information before sending a signal to another element of the network.

Time delays can be of two types. If the current state of a network element depends on the state of another element at only a single arbitrary point in time, we say that this interaction has a *single* delay. If this interaction depends on a number of previous points in time, we refer to it as a *multiple* delay.

It is well known that by adding or removing time delays one can destabilize dynamics, i.e., a globally stable dynamical network can become unstable. It is proved in Chap. 3 that the situation changes if one considers a slightly stronger property, which we call *intrinsic global stability* (see Definition 3.2).

A major and unexpected finding in this chapter is that if a network is intrinsically globally stable, then the addition and removal of singular delays to the network's dynamics will not destabilize the network. Moreover, the network remains stable if we remove any of its multiple time delays. This result, in the case of singular delays, allows one to analyze the stability of a time-delayed network in terms of a nondelayed network, which is usually much simpler, especially from a computational point of view.

Additionally, using this theory of time delays, we show that even if a network's dynamics are not explicitly time delayed, the network still has what we refer to as *implicit time delays*. By removing a network's implicit time delays, we are able to construct a lower-dimensional dynamical network. This network, which we refer to as a network restriction, is similar in many ways to the graph reductions introduced in Chap. 2 and can be used to obtain improved estimates of a network's global stability.

In Chap. 3, we begin by considering the fundamental theorem known as Abel's impossibility theorem (or the Abel–Ruffini theorem). This theorem states that there is no general solution for polynomial equations of degree five or higher, i.e., there is no algebraic formula that represents the roots of a general polynomial of degree $n > 4$. Therefore, there is no algebraic formula for the eigenvalues of $n \times n$ matrices for $n > 4$.

However, interest in practical problems, such as the stability of wave motion, has stimulated development in a branch of linear algebra that deals with estimating the spectra of matrices. This is built on the fundamental theorem of algebra, which states that an $n \times n$ matrix with complex entries has exactly n eigenvalues, including multiplicities, in the complex plane. The goal in this particular theory is to find regions in the complex plane that contain the spectrum of a given matrix.

In Chap. 4, we demonstrate that by combining our method of isospectral reduction with any of the classical methods of eigenvalue estimation, we obtain better estimates than can be obtained using a particular classical method by itself [8]. In this context, better estimates mean that the regions on the complex plane achieved with the help of isospectral transformations are smaller than those obtained without the use of these reductions.

One reason for this improvement is that our method uses more information about the structure of the corresponding matrix. Analogous results are obtained for estimating spectra of the combinatorial and normalized graph Laplacians. We also show that the estimates of a network's spectra can be improved if we have some specific information regarding the network's structure (topology), e.g., some local information about a large graph.

Chapter 5 deals with the pseudospectra and inverse pseudospectra of matrices with complex entries. The pseudospectrum of a complex-valued matrix is a collection of numbers that are sufficiently close (within a *tolerance* $\epsilon > 0$) to the eigenvalues of the matrix. In this chapter, we first extend the definition of pseudospectra to matrices with entries that are rational functions. Since this type of matrix also has a well-defined inverse spectrum, we also introduce the idea of an inverse pseudospectrum for these matrices and study the properties of these sets.

In particular, we show that the pseudospectrum of the reduced matrix is less susceptible to perturbations than the pseudospectrum of the original matrix [28].

Linear mass–spring networks are a major example of applying the techniques of pseudospectra and inverse pseudospectra estimates in this chapter. These networks also allow us to give a physical interpretation of pseudospectra and isospectral reductions.

The final chapter, Chap. 6, deals with yet another application of isospectral transformations. Here, we consider open chaotic dynamical systems, that have a finite Markov partition. In these systems, we suppose that some element (or union of elements) of the Markov partition acts as a "hole." In this case, an orbit that hits the hole stays there forever.

A major characteristic of the dynamics in such systems is the system's survival probability. An open system's survival probability is the probability that an orbit avoids falling into the hole until some fixed point in time. We show that by transforming the underlying open dynamical system using our theory of isospectral transformations, we can obtain improved estimates of a system's survival probability [11].

Overall, the current state of our theory of isospectral transformations is that in each setting in which it has been applied, this theory has led to either improved or entirely new results. Therefore, we are quite optimistic about its future applications. In particular, as this book demonstrates, the theory of isospectral network reductions is ripe for applications to real-life networks, and we hope that this book will help to stimulate such investigations.

Contents

1 Isospectral Matrix Reductions 1
   1.1 Matrices with Rational Function Entries 2
   1.2 Isospectral Matrix Reductions 5
   1.3 Sequential Reductions 12
   1.4 Spectral Inverse 15

2 Dynamical Networks and Isospectral Graph Reductions 19
   2.1 Dynamical Networks as Graphs 20
   2.2 Isospectral Graph Reductions 22
   2.3 Sequential Graph Reductions 31
   2.4 Equivalence Relations 36
   2.5 Weight-Preserving Isospectral Transformations 39
          2.5.1 Branch Expansions 42
   2.6 Isospectral Graph Transformations over Modified Weight Sets 45
          2.6.1 Branch Reweighting 46
          2.6.2 Branch Merging 47

3 Stability of Dynamical Networks 53
   3.1 Networks as Dynamical Systems 54
   3.2 Time-Delayed Dynamical Networks 63
   3.3 Graph Structure of a Dynamical Network 71
   3.4 Implicit Delays and Restrictions of Dynamical Networks 80

4 Improved Eigenvalue Estimates 91
   4.1 Gershgorin-Type Regions 92
   4.2 Brauer-Type Regions 104
   4.3 Brualdi-Type Regions 111
   4.4 Some Applications 123

5	**Pseudospectra and Inverse Pseudospectra**	129
	5.1 Pseudospectra	129
	5.2 Pseudospectra Under Isospectral Reduction	136
	5.3 Inverse Pseudoeigenvalues	139
	5.4 Eigenvalue Inclusions and Equivalence of Definitions	144
6	**Improved Estimates of Survival Probabilities**	147
	6.1 Open Dynamical Systems	148
	6.2 Piecewise Linear Functions	149
	6.3 Nonlinear Estimates	159
	6.4 Improved Escape Estimates	163
References		171
Index		173

Chapter 1
Isospectral Matrix Reductions

The main object of study in this chapter is not networks, which are the main focus of this book, but matrices. The reason is that although a network has more structure than its adjacency matrix, the adjacency matrix of a network is a very convenient and compact way of storing this structural information. We therefore postpone our analysis of a network's graph structure until the next chapter.

In the preface, we stated that one of the major goals of this book is to develop a set of mathematical tools that will allow us to reduce the size of a network while preserving all information relative to its spectrum. Since the spectrum of a network is the eigenvalues of its adjacency matrix, we are, in fact, looking for a way to reduce the size of a matrix while maintaining the matrix's set of eigenvalues.

This may be a bit surprising, especially if we consider the fundamental theorem of algebra, which states that an $n \times n$ matrix with complex entries has exactly n eigenvalues (with multiplicities). Consequently, every matrix smaller than A must have fewer than n eigenvalues.

In this chapter, we will develop the basic tools that will allow us to fulfill this seemingly impossible task of reducing a matrix while maintaining its spectrum. The key is to consider matrices with entries that are not complex numbers but rather functions of a spectral parameter. For our purposes, the class of functions we will use is that of rational functions.

One of our main results in this chapter is to show that a matrix can be reduced in size to a smaller matrix with rational function entries in a way that essentially preserves its eigenvalues. This procedure, called isospectral matrix reduction, reduces a matrix over one of its principal submatrices. We show that a matrix can be reduced over any of its principal submatrices and that it is possible to give a formula for the reduced matrix.

Once we have reduced a matrix, a natural question is whether we can take the reduced matrix and reduce it a second time. The second half of this chapter deals with the question whether a matrix can be sequentially reduced. Here, we prove a

© Springer Science+Business Media New York 2014
L. Bunimovich, B. Webb, *Isospectral Transformations: A New Approach to Analyzing Multidimensional Systems and Networks*, Springer Monographs in Mathematics, DOI 10.1007/978-1-4939-1375-6_1

crucial result, that a matrix can be sequentially reduced and that the matrix resulting from a sequence of reductions does not depend on the particular sequence but only on the final submatrix over which it is reduced.

Aside from the basic theory of isospectral matrix reductions, we also introduce some concepts in this chapter that will be of use later in the book. As is true for the entire book, the practically oriented reader can skip the proofs and simply apply the formulas that are given to analyze a matrix or set of matrices that are of interest in any area of application. In this regard, we note that the computations required to find an isospectral matrix reduction are simple enough that they can be carried out with the aid of any standard software.

1.1 Matrices with Rational Function Entries

As previously stated, one of our major goals is to be able to reduce the size of a matrix in such a way that we preserve the matrix's spectrum, including the multiplicity of each eigenvalue. However, this presupposes that such reductions are possible.

If the matrix A is in $\mathbb{C}^{n \times n}$, then the *eigenvalues* of A are defined as the solutions of its characteristic equation

$$\sigma(A) = \{\lambda \in \mathbb{C} : \det(A - \lambda I) = 0\}.$$

Since the *characteristic polynomial* $\det(A - \lambda I)$ of A has degree n, the fundamental theorem of algebra states that A has exactly n eigenvalues. Hence, if $B \in \mathbb{C}^{m \times m}$, where $m < n$, then A and B cannot have the same spectrum. This alone seems to imply that it is impossible to reduce the size of a matrix while preserving its eigenvalues. This conclusion is certainly true if we limit ourselves to matrices with scalar entries. The only remaining possibility, then, is to consider matrices whose entries are not complex numbers but some other mathematical object, presumably one that carries more information.

For the purposes of reducing a matrix, the class of entries we consider are that of rational functions of λ. Specifically, let $\mathbb{C}[\lambda]$ be the set of polynomials in the complex variable λ with complex coefficients. We denote by $\mathbb{W}$ the set of rational functions of the form

$$\omega(\lambda) = p(\lambda)/q(\lambda),$$

where $p(\lambda), q(\lambda) \in \mathbb{C}[\lambda]$ are polynomials having no common linear factors, i.e., no common roots, and where $q(\lambda)$ is not identically zero.

Each rational function $\omega(\lambda) \in \mathbb{W}$ can be expressed in the form

$$\omega(\lambda) = \frac{a_i \lambda^i + a_{i-1} \lambda^{i-1} + \cdots + a_0}{b_j \lambda^j + b_{j-1} \lambda^{j-1} + \cdots + b_0},$$

1.1 Matrices with Rational Function Entries

where, without loss in generality, we take $b_j = 1$. The domain of $\omega(\lambda)$ consists of all but a finite number of complex numbers for which the polynomial $q(\lambda) = b_j \lambda^j + b_{j-1} \lambda^{j-1} + \cdots + b_0$ is zero.

Addition and multiplication on the set $\mathbb{W}$ are defined as follows. For $p(\lambda)/q(\lambda)$ and $r(\lambda)/s(\lambda)$ in $\mathbb{W}$, let

$$\left(\frac{p}{q} + \frac{r}{s}\right)(\lambda) = \frac{p(\lambda)s(\lambda) + q(\lambda)r(\lambda)}{q(\lambda)s(\lambda)}; \text{ and} \tag{1.1}$$

$$\left(\frac{p}{q} \cdot \frac{r}{s}\right)(\lambda) = \frac{p(\lambda)r(\lambda)}{q(\lambda)s(\lambda)}, \tag{1.2}$$

where the common linear factors on the right-hand side of (1.1) and (1.2) are canceled. The set $\mathbb{W}$ is then a field under the operations of addition and multiplication.

Because we are primarily concerned with the eigenvalues of a matrix, which is a set that includes multiplicities, the following will be important. The element α of the set A that includes multiplicities has *multiplicity m* if there are m elements of A equal to α. If $\alpha \in A$ with multiplicity m and $\alpha \in B$ with multiplicity n, then

(i) the *union* $A \cup B$ is a set in which α has multiplicity $m + n$; and
(ii) the *difference* $A - B$ is a set in which α has multiplicity $m - n$ if $m - n > 0$ and where $\alpha \notin A - B$ otherwise.

Definition 1.1. Let $\mathbb{W}^{n \times n}$ denote the set of $n \times n$ matrices with entries in $\mathbb{W}$. For a matrix $M(\lambda) \in \mathbb{W}^{n \times n}$, the determinant $\det\bigl(M(\lambda) - \lambda I\bigr)$ is given by

$$\det\bigl(M(\lambda) - \lambda I\bigr) = p(\lambda)/q(\lambda) \tag{1.3}$$

for some $p(\lambda)/q(\lambda) \in \mathbb{W}$. The *spectrum* (or set of eigenvalues) of $M(\lambda)$ is the set

$$\sigma(M) = \{\lambda \in \mathbb{C} : p(\lambda) = 0\}.$$

The *inverse spectrum* (or inverse eigenvalues) of $M(\lambda)$ is the set

$$\sigma^{-1}(M) = \{\lambda \in \mathbb{C} : q(\lambda) = 0\}.$$

Both $\sigma(M)$ and $\sigma^{-1}(M)$ are understood to be sets that include multiplicities. For example, if the polynomial $p(\lambda) \in \mathbb{C}[\lambda]$ in (1.3) factors as

$$p(\lambda) = \prod_{i=1}^{m} (\lambda - \alpha_i)^{n_i} \text{ for } \alpha_i \in \mathbb{C} \text{ and } n_i \in \mathbb{N},$$

then $\{\lambda \in \mathbb{C} : p(\lambda) = 0\}$ is the set in which α_i has multiplicity n_i.

Example 1.1. Consider the matrix $M(\lambda) \in \mathbb{W}^{4\times 4}$ given by

$$M(\lambda) = \begin{bmatrix} 2\lambda+2 & \frac{1}{\lambda} & 0 & \frac{1}{\lambda} \\ \frac{1}{\lambda} & 2\lambda+2 & \frac{1}{\lambda} & 0 \\ 0 & \frac{1}{\lambda} & 2\lambda+2 & \frac{1}{\lambda} \\ \frac{1}{\lambda} & 0 & \frac{1}{\lambda} & 2\lambda+2 \end{bmatrix}. \quad (1.4)$$

As one can compute,

$$\det(M(\lambda)-\lambda I) = \frac{(\lambda+2)^2(\lambda^2(\lambda+2)^2-4)}{\lambda^2},$$

implying $\sigma(M) = \{-2, -2, -1\pm i, -1\pm\sqrt{3}\}$. Here we note that although M is a 4×4 matrix, it has six eigenvalues including multiplicities. This gives us our first example of a matrix with more eigenvalues than either rows or columns.

Example 1.1 suggests that it may be possible to reduce the size of a network (or matrix) while at the same time preserving its spectrum. For instance, the matrix

$$A = \begin{bmatrix} -2 & 1 & 1 & 0 & 0 & 0 \\ 0 & -1+i & 0 & 1 & 0 & 0 \\ 0 & 0 & -1-i & 0 & 1 & 0 \\ 0 & 0 & 0 & -1-\sqrt{3} & 0 & 1 \\ 0 & 0 & 0 & 0 & -1+\sqrt{3} & 1 \\ 0 & 0 & 0 & 0 & 0 & -2 \end{bmatrix} \quad (1.5)$$

and the matrix $M(\lambda)$ in Example 1.1 have the same spectrum. However, it is not at all obvious whether some procedure exists that would allows us to reduce the matrix A to the smaller matrix $M(\lambda)$ or what such a procedure might be.

Since $\mathbb{C} \subset \mathbb{W}$, we note that Definition 1.1 is an extension of the standard definition of the eigenvalues of a matrix to the matrices $\mathbb{W}^{n\times n}$. In particular, if the matrix $M(\lambda)$ is in $\mathbb{C}^{n\times n}$, then $\sigma(M)$ are the standard eigenvalues of M.

Because our motivation is to extend the spectral theory of matrices, our standard practice will be to take concepts used in the context of scalar-valued matrices and demonstrate that they can be applied to $\mathbb{W}^{n\times n}$.

For instance, the matrix $A(\lambda) \in \mathbb{W}^{n\times n}$ is said to be *invertible* if there is a matrix $B(\lambda) \in \mathbb{W}^{n\times n}$ such that $A(\lambda)B(\lambda) = I$ is the $n\times n$ identity matrix. If $A(\lambda)$ is invertible, we use the standard notation $A(\lambda)^{-1}$ to denote its inverse. As an example, if

$$M(\lambda) = \begin{bmatrix} \frac{1}{\lambda} & 1 \\ 0 & \lambda \end{bmatrix} \text{ then } M(\lambda)^{-1} = \begin{bmatrix} \lambda & -1 \\ 0 & \frac{1}{\lambda} \end{bmatrix}.$$

In what follows, we may, for convenience, suppress the dependence of the matrix $M(\lambda) \in \mathbb{W}^{n \times n}$ on λ and simply write M. One reason for this is that for much of the theory that is developed here, we do not evaluate $M(\lambda)$ at any particular point $\lambda \in \mathbb{C}$. Rather, we consider M formally as a matrix with rational function entries and not as a function of the spectral parameter λ.

However, when we do consider the matrix $M(\lambda) \in \mathbb{W}^{n \times n}$ to be a function of λ, we mean that M is the function

$$M : \mathrm{dom}(M) \to \mathbb{C}^{n \times n},$$

where $\mathrm{dom}(M)$ is the set of all but the finite number of complex numbers for which each entry of $M(\lambda)$ is defined. Surprisingly, it may be the case that $\sigma(M)$ is not a subset of $\mathrm{dom}(M)$, as the following example shows.

Example 1.2. Consider the matrix $M(\lambda) \in \mathbb{W}^{2 \times 2}$ given by

$$M(\lambda) = \begin{bmatrix} \lambda_0 & (\lambda - \lambda_0)^{-1} \\ 0 & \lambda_0 \end{bmatrix},$$

where $\lambda_0 \in \mathbb{C}$. As one can compute, $\det(M(\lambda) - \lambda I) = (\lambda - \lambda_0)^2$, implying that $\sigma(M) = \{\lambda_0, \lambda_0\}$. Therefore, $\sigma(M)$ is not a subset of $\mathrm{dom}(M) = \mathbb{C} - \{\lambda_0\}$.

1.2 Isospectral Matrix Reductions

Having introduced the notions of a spectrum and inverse spectrum of a matrix with rational function entries, we can now describe an *isospectral reduction* of a matrix $M \in \mathbb{W}^{n \times n}$. The major goals in this section are first, to describe the process of isospectral reduction, and second, to compare the spectrum of a reduced matrix with that of the original unreduced matrix.

For $M \in \mathbb{W}^{n \times n}$, let $N = \{1, \ldots, n\}$. If the sets $R, C \subseteq N$ are nonempty, we denote by M_{RC} the $|R| \times |C|$ *submatrix* of M with rows indexed by R and columns by C. Suppose the nonempty sets $S \subset N$ and its complement $\bar{S} = N - S$ are nonempty. The *Schur complement* $M/M_{\bar{S}\bar{S}} \in \mathbb{W}^{|S| \times |S|}$ of $M_{\bar{S}\bar{S}}$ in M is the matrix

$$M/M_{\bar{S}\bar{S}} = M_{SS} - M_{S\bar{S}} M_{\bar{S}\bar{S}}^{-1} M_{\bar{S}S}, \tag{1.6}$$

assuming that the submatrix $M_{\bar{S}\bar{S}}$ is invertible.

The Schur complement arises in many applications (see, e.g., [18]). For our purposes, the Schur complement allows us to define the reduction of a matrix $M \in \mathbb{W}^{n \times n}$.

Definition 1.2. For $M(\lambda) \in \mathbb{W}^{n \times n}$, let S and $\bar{S}$ form a nonempty partition of N. The *isospectral reduction* of M over the set S is the matrix

$$\mathcal{R}_\lambda(M; S) = M_{SS} - M_{S\bar{S}}(M_{\bar{S}\bar{S}} - \lambda I)^{-1} M_{\bar{S}S} \in \mathbb{W}^{|S| \times |S|} \quad (1.7)$$

if the matrix $M_{\bar{S}\bar{S}} - \lambda I$ is invertible.

As can be seen from Definition 1.2, the reduced matrix $\mathcal{R}(M; S)$ exists if and only if the matrix $M_{\bar{S}\bar{S}} - \lambda I$ is invertible. Moreover, $\mathcal{R}_\lambda(M; S)$ is a Schur complement plus a multiple of the identity:

$$\mathcal{R}_\lambda(M; S) = (M - \lambda I)/(M_{\bar{S}\bar{S}} - \lambda I) + \lambda I. \quad (1.8)$$

In what follows, we will more often than not suppress the dependence of the reduced matrix $\mathcal{R}_\lambda(M; S)$ on λ and instead write it as $\mathcal{R}(M; S)$.

Example 1.3. Consider the matrix $M \in \mathbb{W}^{6 \times 6}$ with $(0, 1)$-entries given by

$$M = \begin{bmatrix} 0 & 0 & 1 & 1 & 0 & 0 \\ 0 & 1 & 0 & 0 & 1 & 1 \\ 1 & 0 & 1 & 0 & 0 & 0 \\ 0 & 1 & 0 & 1 & 0 & 0 \\ 1 & 0 & 0 & 0 & 0 & 0 \\ 0 & 1 & 0 & 0 & 0 & 0 \end{bmatrix}.$$

For $S = \{1, 2\}$ and $\bar{S} = \{3, 4, 5, 6\}$, one can compute that

$$(M_{\bar{S}\bar{S}} - \lambda I)^{-1} = \begin{bmatrix} \frac{1}{1-\lambda} & 0 & 0 & 0 \\ 0 & \frac{1}{1-\lambda} & 0 & 0 \\ 0 & 0 & -\frac{1}{\lambda} & 0 \\ 0 & 0 & 0 & -\frac{1}{\lambda} \end{bmatrix}.$$

The isospectral reduction of M over $S = \{1, 2\}$ is then

$$\mathcal{R}(M; S) = \begin{bmatrix} 0 & 0 \\ 0 & 1 \end{bmatrix} - \begin{bmatrix} 1 & 1 & 0 & 0 \\ 0 & 0 & 1 & 1 \end{bmatrix} \begin{bmatrix} \frac{1}{1-\lambda} & 0 & 0 & 0 \\ 0 & \frac{1}{1-\lambda} & 0 & 0 \\ 0 & 0 & -\frac{1}{\lambda} & 0 \\ 0 & 0 & 0 & -\frac{1}{\lambda} \end{bmatrix} \begin{bmatrix} 1 & 0 \\ 0 & 1 \\ 1 & 0 \\ 0 & 1 \end{bmatrix}$$

$$= \begin{bmatrix} \frac{1}{\lambda-1} & \frac{1}{\lambda-1} \\ \frac{1}{\lambda} & \frac{\lambda+1}{\lambda} \end{bmatrix} \in \mathbb{W}^{2 \times 2}.$$

If a matrix has an isospectral reduction, the spectrum and inverse spectrum of the isospectral reduction and the original matrix are related in the following way.

1.2 Isospectral Matrix Reductions

Theorem 1.1 (Spectrum and Inverse Spectrum of Isospectral Reductions).
For $M(\lambda) \in \mathbb{W}^{n \times n}$, let S and $\bar{S}$ form a nonempty partition of $N = \{1, \ldots, n\}$. If $\mathscr{R}_\lambda(M; S)$ exists, then its spectrum and inverse spectrum are given by

$$\sigma\big(\mathscr{R}(M; S)\big) = \big(\sigma(M) \cup \sigma^{-1}(M_{\bar{S}\bar{S}})\big) - \big(\sigma(M_{\bar{S}\bar{S}}) \cup \sigma^{-1}(M)\big); \text{ and}$$

$$\sigma^{-1}\big(\mathscr{R}(M; S)\big) = \big(\sigma(M_{\bar{S}\bar{S}}) \cup \sigma^{-1}(M)\big) - \big(\sigma(M) \cup \sigma^{-1}(M_{\bar{S}\bar{S}})\big).$$

Proof. For $M \in \mathbb{W}^{n \times n}$, we may assume, without loss of generality, that M has the block matrix form

$$M = \begin{bmatrix} M_{\bar{S}\bar{S}} & M_{\bar{S}S} \\ M_{S\bar{S}} & M_{SS} \end{bmatrix}, \tag{1.9}$$

where $M_{\bar{S}\bar{S}} - \lambda I$ is invertible.

Note that the determinant of a matrix and that of its Schur complement are related by the identity

$$\det \begin{bmatrix} A & B \\ C & D \end{bmatrix} = \det(A) \cdot \det(D - CA^{-1}B), \tag{1.10}$$

provided that the submatrix A is invertible. Using this identity on the matrix $M - \lambda I$ yields

$$\det(M - \lambda I) = \det(M_{\bar{S}\bar{S}} - \lambda I) \cdot \det\big((M_{SS} - \lambda I) - M_{S\bar{S}}(M_{\bar{S}\bar{S}} - \lambda I)^{-1}M_{\bar{S}S}\big).$$

Therefore,

$$\det\big(\mathscr{R}(M; S) - \lambda I\big) = \frac{\det(M - \lambda I)}{\det(M_{\bar{S}\bar{S}} - \lambda I)}. \tag{1.11}$$

To compare the eigenvalues of $\mathscr{R}(M; S)$, M, and $M_{\bar{S}\bar{S}}$, write

$$\det(M - \lambda I) = \frac{p(\lambda)}{q(\lambda)} \text{ and } \det(M_{\bar{S}\bar{S}} - \lambda I) = \frac{t(\lambda)}{u(\lambda)},$$

for some $p(\lambda)/q(\lambda), t(\lambda)/u(\lambda) \in \mathbb{W}$. Hence

$$\det\big(\mathscr{R}(M; S) - \lambda I\big) = \frac{p(\lambda)u(\lambda)}{q(\lambda)t(\lambda)}.$$

Let $P = \{\lambda \in \mathbb{C} : p(\lambda) = 0\}$, $Q = \{\lambda \in \mathbb{C} : q(\lambda) = 0\}$, $T = \{\lambda \in \mathbb{C} : t(\lambda) = 0\}$, and $U = \{\lambda \in \mathbb{C} : u(\lambda) = 0\}$ be the sets that include multiplicities. By canceling common linear factors, Definition 1.1 implies

$$\sigma\bigl(\mathscr{R}(M;S)\bigr) = \{\lambda \in \mathbb{C} : p(\lambda)u(\lambda) = 0\} - \{\lambda \in \mathbb{C} : q(\lambda)t(\lambda) = 0\}$$
$$= (P \cup U) - (Q \cup T); \text{ and}$$
$$\sigma^{-1}\bigl(\mathscr{R}(M;S)\bigr) = \{\lambda \in \mathbb{C} : q(\lambda)t(\lambda) = 0\} - \{\lambda \in \mathbb{C} : p(\lambda)u(\lambda) = 0\}$$
$$= (Q \cup T) - (P \cup U).$$

Since $P = \sigma(M)$, $Q = \sigma(M_{\bar{S}\bar{S}})$, $T = \sigma^{-1}(M)$, and $U = \sigma^{-1}(M_{\bar{S}\bar{S}})$ the result follows. □

Since a matrix $M \in \mathbb{C}^{n \times n}$ has no inverse spectrum (i.e., $\sigma^{-1}(M) = \emptyset$), Theorem 1.1 applied to a complex-valued matrix has the following corollary.

Corollary 1.1. *For $M \in \mathbb{C}^{n \times n}$, let S and $\bar{S}$ form a nonempty partition of N. Then*

(i) $\sigma\bigl(\mathscr{R}(M;S)\bigr) = \sigma(M) - \sigma(M_{\bar{S}\bar{S}})$; *and*
(ii) $\sigma^{-1}\bigl(\mathscr{R}(M;S)\bigr) = \sigma(M_{\bar{S}\bar{S}}) - \sigma(M)$.

Example 1.4. Let M and S be as in Example 1.3. As one can compute, the spectrum and inverse spectrum of M are given by $\sigma(M) = \{2, -1, 1, 1, 0, 0\}$ and $\sigma(M_{\bar{S}\bar{S}}) = \{1, 1, 0, 0\}$. From Corollary 1.1, we have

$$\sigma\bigl(\mathscr{R}(M;S)\bigr) = \{2, -1, 1, 1, 0, 0\} - \{1, 1, 0, 0\} = \{2, -1\}; \text{ and}$$
$$\sigma^{-1}\bigl(\mathscr{R}(M;S)\bigr) = \{1, 1, 0, 0\} - \{2, -1, 1, 1, 0, 0\} = \emptyset.$$

Observe that by reducing M over S, we lose all eigenvalues in the spectrum of the submatrix $\sigma(M_{\bar{S}\bar{S}}) = \{1, 1, 0, 0\}$.

Theorem 1.1 describes exactly which eigenvalues we gain from an isospectral reduction and which we may lose. Specifically, by isospectrally reducing the matrix M over S, we always gain the eigenvalues $\sigma^{-1}(M_{\bar{S}\bar{S}})$ and lose all eigenvalues in the set $\sigma(M_{\bar{S}\bar{S}}) \cup \sigma^{-1}(M)$.

In this sense, an isospectral reduction of a matrix preserves the spectral information of the original matrix. However, it may not always be possible to reduce a matrix $M \in \mathbb{W}^{n \times n}$ over a particular set $S \subset N$.

Example 1.5. Consider the matrix $M \in \mathbb{W}^{2 \times 2}$ given by

$$M = \begin{bmatrix} 1 & 1 \\ 1 & \lambda \end{bmatrix}. \tag{1.12}$$

For $S = \{1\}$ and $\bar{S} = \{2\}$, note that $M_{\bar{S}\bar{S}} - \lambda I = [0]$, which is not invertible. Therefore, M cannot be isospectrally reduced over S.

In general, there is no way to know beforehand whether the isospectral reduction $\mathscr{R}(M;S)$ exists without attempting to compute $(M_{\bar{S}\bar{S}} - \lambda I)^{-1}$. However, the following subset of $\mathbb{W}^{n \times n}$ can always be reduced over every nonempty subset $S \subset N$.

1.2 Isospectral Matrix Reductions

For $p(\lambda) \in \mathbb{C}[\lambda]$, let $\deg(p)$ denote the degree of the polynomial $p(\lambda)$. If the rational function $w(\lambda)$ is equal to $p(\lambda)/q(\lambda)$, where $p(\lambda), q(\lambda) \in \mathbb{C}[\lambda]$ and $p(\lambda) \neq 0$, we define the *degree* of the rational function $w(\lambda)$ by

$$\pi(w) = \deg(p) - \deg(q).$$

When $p(\lambda) = 0$, we let $\pi(w) = 0$.

Definition 1.3. Let $\mathbb{W}_\pi$ be the set of rational functions

$$\mathbb{W}_\pi = \{w(\lambda) \in \mathbb{W} : \pi(w) \leq 0\},$$

and let $\mathbb{W}_\pi^{n \times n}$ be the set of $n \times n$ matrices with entries in $\mathbb{W}_\pi$.

The set $\mathbb{W}_\pi \subset \mathbb{W}$ consists of the rational functions for which the degree of the numerator is less than or equal to the degree of the denominator. To describe why this collection of rational functions is so useful in the theory of isospectral reductions, we begin by proving the following statement.

Lemma 1.1. *Suppose $\omega_i(\lambda) = p_i(\lambda)/q_i(\lambda)$, where $p_i(\lambda), q_i(\lambda) \in \mathbb{C}[\lambda]$ and $q_i(\lambda)$ is nonzero for $1 \leq i \leq n$. For $1 \leq i, j \leq n$, the following properties hold:*

$$\pi\left(\sum_{i=1}^n \omega_i\right) = \max_{1 \leq i \leq n}\{\pi(\omega_i) : \omega_i \neq 0\}; \tag{1.13}$$

$$\pi\left(\prod_{i=1}^n \omega_i\right) = \begin{cases} \sum_{i=1}^n \pi(\omega_i) & \text{if } \forall\, i \in \{1, \cdots, n\}\ \omega_i \neq 0 \\ 0 & \text{otherwise}; \end{cases} \tag{1.14}$$

$$\pi(\omega_i/\omega_j) = \begin{cases} \pi(\omega_i) - \pi(\omega_j) & \text{if } \omega_i \neq 0 \\ 0 & \text{otherwise} \end{cases} \text{ for } \omega_j \neq 0;\text{ and} \tag{1.15}$$

$$\pi(\omega_i - \lambda) = 1 \text{ for } \omega_i(\lambda) \in \mathbb{W}_\pi. \tag{1.16}$$

Proof. We first prove equations (1.13) and (1.14) for $n = 2$. The general statements follow by induction. For $\omega_i(\lambda) = p_i(\lambda)/q_i(\lambda)$ and $i = 1, 2$, we have

$$\pi(\omega_1 + \omega_2) = \pi\left(\frac{p_1 q_2 + p_2 q_1}{q_1 q_2}\right)$$
$$= \max\{\deg(p_1 q_2), \deg(p_2 q_1)\} - \deg(q_1) - \deg(q_2)$$
$$= \max\{\pi(\omega_1), \pi(\omega_2))\}.$$

Similarly, the product $\omega_1(\lambda)\omega_2(\lambda)$ has degree

$$\pi(\omega_1 + \omega_2) = \pi\left(\frac{p_1 p_2}{q_1 q_2}\right)$$
$$= \deg(p_1) + \deg(p_2) - \deg(q_1) - \deg(q_2)$$
$$= \pi(\omega_1) + \pi(\omega_2).$$

To prove equation (1.15), we observe that for $\omega_i(\lambda), \omega_j(\lambda) \neq 0$, we have

$$\pi(\omega_i/\omega_j) = \pi\left(\frac{p_i q_j}{q_i q_j}\right)$$
$$= \deg(p_i) + \deg(q_j) - \deg(q_i) - \deg(p_j)$$
$$= \pi(\omega_i) - \pi(\omega_j).$$

If $\omega_i(\lambda) = 0$, then $\omega_i/\omega_j = 0$, implying $\pi(\omega_i/\omega_j) = 0$.

For equation (1.15), we have

$$\pi(\omega_i - \lambda) = \pi\left(\frac{p_i - \lambda q_i}{q_i}\right)$$
$$= \max\{\deg(p_i), \deg(q_i) + 1\} - \deg(q_i) = 1,$$

because $\deg(p_i) \geq \deg(q_i)$, since $\omega_i(\lambda) \in \mathbb{W}_\pi$.

Equations (1.13) and (1.14) directly imply that $\mathbb{W}_\pi$ is closed under addition and multiplication. However, $\mathbb{W}_\pi$ is not a field, since most elements in this set do not have a multiplicative inverse.

Theorem 1.2 (Existence of Isospectral Reductions). *Let $M(\lambda) \in \mathbb{W}_\pi^{n \times n}$. If S and $\bar{S}$ form a nontrivial partition of N, then $\mathcal{R}(M; S)$ exists and is in $\mathbb{W}_\pi^{|S| \times |S|}$.*

Proof. Let $M \in \mathbb{W}_\pi^{n \times n}$. The inverse of the matrix $M - \lambda I$ can be written as

$$(M - \lambda I)^{-1} = \frac{1}{\det(M - \lambda I)} \operatorname{adj}(M - \lambda I), \qquad (1.17)$$

where $\operatorname{adj}(M - \lambda I)$ is the adjugate matrix of $M - \lambda I$, i.e., the matrix with entries

$$\operatorname{adj}(M - \lambda I)_{ij} = (-1)^{i+j} \det(\mathcal{M}_{ji}), \ 1 \leq i, j \leq n, \qquad (1.18)$$

where $\mathcal{M}_{ij} \in \mathbb{W}^{(n-1) \times (n-1)}$ is obtained by deleting the ith row and jth column of $M - \lambda I$.

1.2 Isospectral Matrix Reductions

Additionally, the determinant of $M - \lambda I$ can be written as

$$\det(M - \lambda I) = \sum_{\rho \in \mathscr{P}_n} \left(\text{sgn}(\rho) \prod_{i=1}^{n} (M - \lambda I)_{i,\rho(i)} \right), \quad (1.19)$$

where the sum is taken over the set $\mathscr{P}_n$ of permutations on N. The sign $\text{sgn}(\rho)$ of the permutation $\rho \in \mathscr{P}_n$ is 1 (respectively -1) if ρ is the composition of an even (respectively odd) number of permutations of two elements.

Using (1.14) and (1.16), we see that the term in (1.19) corresponding to the identity permutation $\rho = \text{id} \in \mathscr{P}_n$ has degree n, while for $\rho \neq \text{id}$, the other terms have degree strictly smaller than n. Equation (1.13) then implies that

$$\pi\big(\det(M - \lambda I)\big) = n. \quad (1.20)$$

Therefore, $\det(M - \lambda I)$ is not identically zero, implying via equation (1.17) that the inverse $(M - \lambda I)^{-1}$ exists. Similarly, for $i \in N$, the matrix $\mathscr{M}_{ii}$ is equal to $\tilde{\mathscr{M}}_{ii} - \lambda I$ for some $\tilde{\mathscr{M}} \in \mathbb{W}_\pi^{(n-1) \times (n-1)}$. Hence,

$$\pi\big(\det(\mathscr{M}_{ii})\big) = n - 1, \text{ for } i \in N. \quad (1.21)$$

For $i \neq j$, the matrices $\mathscr{M}_{ij} \in \mathbb{W}^{(n-1) \times (n-1)}$ contain $n - 2$ entries of the form $M_{k\ell} - \lambda$, while all other entries of $\mathscr{M}_{ij}$ belong to the set $\mathbb{W}_\pi$. Hence, equations (1.14) and (1.16) imply that for $i \neq j$,

$$\pi\big(\det(\mathscr{M}_{ij})\big) \leq n - 2, \text{ for } i, j \in N, \quad (1.22)$$

since for $\rho \in \mathscr{P}_{n-1}$, at most $n - 2$ terms in the product $\prod_{k=1}^{n-1}(\mathscr{M}_{ij})_{k,\rho(k)}$ have the form $M_{k\ell} - \lambda$.

Given that the degree of $\det(\mathscr{M}_{ij})$ in (1.22) may be zero, equations (1.20)–(1.22) together with (1.15) imply that $\pi((M - \lambda I)_{ij}^{-1}) \leq 0$ for all $1 \leq i, j \leq n$. Therefore, for every $M \in \mathbb{W}_\pi^{n \times n}$, the matrix $M - \lambda I$ is invertible, and $(M - \lambda I)^{-1} \in \mathbb{W}_\pi^{n \times n}$.

Now suppose that S and $\bar{S}$ form a nontrivial partition of N. Since $M_{\bar{S}\bar{S}} \in \mathbb{W}_\pi^{|\bar{S}| \times |\bar{S}|}$, it follows that $(M_{\bar{S}\bar{S}} - \lambda I)^{-1} \in \mathbb{W}_\pi^{|\bar{S}| \times |\bar{S}|}$. Definition 1.2 along with (1.14) and (1.16) then implies that $\mathscr{R}(M; S)$ exists and has entries in $\mathbb{W}_\pi$. $\square$

Note that Theorem 1.2 implies that the reduction $\mathscr{R}(M; S)$ exists for every matrix $M \in \mathbb{W}_\pi^{n \times n}$ and $S \subset N$. The reason then, that we were unable to reduce the matrix M in Example 1.5 is that M did not belong to $\mathbb{W}_\pi^{2 \times 2}$. Therefore, Theorem 1.2 does not apply.

Although Theorem 1.2 does not apply to every possible matrix, it does apply to those we consider most often, namely, those matrices that have complex-valued entries. This we summarize in the following remark.

Remark 1.1. If $c \in \mathbb{C}$, then $c = c/1$ has degree $\pi(c) = 0$. As a consequence, the complex-valued matrices $\mathbb{C}^{n \times n}$ are contained in $\mathbb{W}_\pi^{n \times n}$. Theorem 1.2 therefore implies that every complex-valued matrix $A \in \mathbb{C}^{n \times n}$ can be reduced over every nonempty set $S \subset N$.

1.3 Sequential Reductions

In the previous section, we observed that the isospectral reduction $\mathcal{R}(M; S)$ of a matrix $M \in \mathbb{W}_\pi^{n \times n}$ is again a matrix in $\mathbb{W}_\pi^{m \times m}$. According to Theorem 1.2, it is therefore possible to reduce the matrix $\mathcal{R}(M; S)$ over some subset of S. That is, we may sequentially reduce every matrix $M \in \mathbb{W}_\pi^{n \times n}$.

A natural question that arises is this: to what extent does a sequentially reduced matrix depend on the particular sequence of index sets over which it has been reduced? As it turns out, if a matrix has been reduced over the index set S_1, and then over S_2, and so on up to the index set S_m, then the resulting matrix depends only on the index set S_m.

To formalize this, let $M \in \mathbb{W}_\pi^{n \times n}$ and suppose there are nonempty sets $S_1, \ldots, S_m$ such that $N \supset S_1 \supset \ldots, \supset S_m$. Then M can be *sequentially reduced* over the sets $S_1, \ldots, S_m$, where we write

$$\mathcal{R}(M; S_1, \ldots, S_m) = \mathcal{R}\bigl(\ldots \mathcal{R}(\mathcal{R}(M; S_1); S_2) \ldots; S_m\bigr)$$

to indicate this sequence of reductions. If M is sequentially reduced over the index sets $S_1, \ldots, S_m$, we call S_m the *final index set* of this sequence of reductions.

Theorem 1.3 (Uniqueness of Sequential Reductions). *For $M(\lambda) \in \mathbb{W}_\pi^{n \times n}$, suppose $N \supset S_1 \supset \ldots, \supset S_m$, where S_m is nonempty. Then*

$$\mathcal{R}(M; S_1, \ldots, S_m) = \mathcal{R}(M; S_m).$$

That is, in a sequence of reductions, the resulting matrix is completely specified by the final index set.

To prove Theorem 1.3 we require the following lemma.

Lemma 1.2. *Let the nonempty sets S, T, and $\overline{S \cup T}$ partition N. If $M(\lambda) \in \mathbb{W}_\pi^{n \times n}$, then $\mathcal{R}(M; S \cup T, S) = \mathcal{R}(M; S)$.*

Proof. Let the nonempty sets S, T, and $U = \overline{S \cup T}$ partition N. We assume, without loss of generality, that $M \in \mathbb{W}_\pi^{n \times n}$ can be written as

$$M = \begin{bmatrix} M_{SS} & M_{ST} & M_{SU} \\ M_{TS} & M_{TT} & M_{TU} \\ M_{US} & M_{UT} & M_{UU} \end{bmatrix}.$$

1.3 Sequential Reductions

Using the definition of isospectral reduction, we have

$$\mathscr{R}(M;S) = M_{SS} - \begin{bmatrix} M_{ST} & M_{SU} \end{bmatrix} \begin{bmatrix} M_{TT} - \lambda I & M_{TU} \\ M_{UT} & M_{UU} - \lambda I \end{bmatrix}^{-1} \begin{bmatrix} M_{TS} \\ M_{US} \end{bmatrix} \text{ and} \tag{1.23}$$

$$\mathscr{R}(M;S \cup T) = \begin{bmatrix} M_{SS} & M_{ST} \\ M_{TS} & M_{TT} \end{bmatrix} - \begin{bmatrix} M_{SU} \\ M_{TU} \end{bmatrix} (M_{UU} - \lambda I)^{-1} \begin{bmatrix} M_{US} & M_{UT} \end{bmatrix}. \tag{1.24}$$

Taking the isospectral reduction of $\mathscr{R}(M;S \cup T)$ over S in (1.24), we obtain

$$\mathscr{R}(M;S \cup T, S)$$
$$= M_{SS} - M_{SU} K(\lambda)^{-1} M_{US}$$
$$- \left[(M_{ST} - M_{SU} K(\lambda)^{-1} M_{UT}) L(\lambda)^{-1} (M_{TS} - M_{TU} K(\lambda)^{-1} M_{US}) \right], \tag{1.25}$$

where $K(\lambda) \equiv M_{UU} - \lambda I$ and $L(\lambda) \equiv M_{TT} - \lambda I - M_{TU} K(\lambda)^{-1} M_{UT}$. Note that both $K(\lambda)^{-1}$ and $L(\lambda)^{-1}$ exist, as can be seen from the proof of Theorem 1.2. To obtain the desired result, we need to verify that expressions (1.23) and (1.25) are equal.

Recall the following identity for the inverse of an invertible square matrix M with 2×2 blocks:

$$M^{-1} = \begin{bmatrix} A & B \\ C & D \end{bmatrix}^{-1} = \begin{bmatrix} E^{-1} & -E^{-1} B D^{-1} \\ -D^{-1} C E^{-1} & D^{-1} + D^{-1} C E^{-1} B D^{-1} \end{bmatrix}, \tag{1.26}$$

where D is an invertible square matrix and $E = A - BD^{-1}C$ is the Schur complement of D in M. We note that if M and D are invertible in (1.26), then the same is true of the matrix E.

From the proof of Theorem 1.2, it follows that the 2×2 block matrix appearing in (1.23) is invertible, as is the submatrix $M_{UU} - \lambda I$. Using (1.26) to find the inverse of this 2×2 block matrix, we get

$$\begin{bmatrix} M_{TT} - \lambda I & M_{TU} \\ M_{UT} & M_{UU} - \lambda I \end{bmatrix}^{-1}$$
$$= \begin{bmatrix} L(\lambda)^{-1} & -L(\lambda)^{-1} M_{TU} K(\lambda)^{-1} \\ -K(\lambda)^{-1} M_{UT} L(\lambda)^{-1} & K(\lambda)^{-1} + K(\lambda)^{-1} M_{UT} L(\lambda)^{-1} M_{TU} K(\lambda)^{-1} \end{bmatrix}. \tag{1.27}$$

Using (1.27) in (1.23), we get (1.25), completing the proof. □

We now give a proof of Theorem 1.3.

Proof. For $M \in \mathbb{W}_\pi^{n \times n}$, suppose $N \subset S_1 \supset \cdots \supset S_m$, where $S_m \neq \emptyset$. If $m = 2$, then Lemma 1.2 directly implies that $\mathscr{R}(M; S_1, S_2) = \mathscr{R}(M; S_2)$.

For $2 \leq k < m$, suppose $\mathscr{R}(M; S_1, \ldots, S_k) = \mathscr{R}(M; S_k)$. Then

$$\mathscr{R}(M; S_1, \ldots, S_k, S_{k+1}) = \mathscr{R}(M; S_k, S_{k+1}) = \mathscr{R}(M; S_{k+1}),$$

where the second equality follows from Lemma 1.2. By induction, it then follows that $\mathscr{R}(M; S_1, \ldots, S_m) = \mathscr{R}(M; S_m)$. □

Example 1.6. Let $M \in \mathbb{C}^{4 \times 4}$ be the matrix with $(0, 1)$-entries given by

$$M = \begin{bmatrix} 1 & 0 & 1 & 0 \\ 0 & 1 & 0 & 1 \\ 0 & 1 & 1 & 1 \\ 1 & 0 & 1 & 1 \end{bmatrix},$$

and let $S = \{1, 2\}$. Our goal in this example is to illustrate that

$$\mathscr{R}(M; S) = \mathscr{R}(M; S \cup \{3\}, S) = \mathscr{R}(M; S \cup \{4\}, S).$$

One can compute

$$\mathscr{R}(M; S \cup \{3\}) = \begin{bmatrix} 1 & 0 & 1 \\ \frac{1}{\lambda-1} & 1 & \frac{1}{\lambda-1} \\ \frac{1}{\lambda-1} & 1 & \frac{\lambda}{\lambda-1} \end{bmatrix} \text{ and } \mathscr{R}(M; S \cup \{4\}) = \begin{bmatrix} 1 & \frac{1}{\lambda-1} & \frac{1}{\lambda-1} \\ 0 & 1 & 1 \\ 1 & \frac{1}{\lambda-1} & \frac{\lambda}{\lambda-1} \end{bmatrix}.$$

Although $\mathscr{R}(M; S \cup \{3\}) \neq \mathscr{R}(M; S \cup \{4\})$, note that by reducing both of these matrices over $S = \{1, 2\}$, one has

$$\mathscr{R}(M; S) = \mathscr{R}(M; S \cup \{3\}, S) = \mathscr{R}(M; S \cup \{4\}, S) = \begin{bmatrix} \frac{\lambda^2 - 2\lambda + 1}{\lambda^2 - 2\lambda} & \frac{\lambda - 1}{\lambda^2 - 2\lambda} \\ \frac{\lambda - 1}{\lambda^2 - 2\lambda} & \frac{\lambda^2 - 2\lambda + 1}{\lambda^2 - 2\lambda} \end{bmatrix}.$$

Additionally, $\sigma(M) = \{\frac{1}{2}(3 \pm \sqrt{5}), \frac{1}{2}(1 \pm \sqrt{-3})\}$ and $\sigma(M_{\bar{S}\bar{S}}) = \{0, 2\}$. Hence, the matrix M and the reduced matrix $\mathscr{R}(M, S)$ have the same eigenvalues by Corollary 1.1.

Example 1.6 illustrates the fact that an isospectral reduction need not have any effect on the spectrum of a matrix.

1.4 Spectral Inverse

In this section, we introduce a matrix transformation that exchanges the spectrum and inverse spectrum of a matrix $M \in \mathbb{W}^{n \times n}$. This transformation will, in fact, be useful in the following chapters, in which we will use it to investigate and define concepts related to the inverse spectrum of a matrix. We introduce this notion here because it is first and foremost a matrix operation.

Definition 1.4. For $M(\lambda) \in \mathbb{W}^{n \times n}$, let $\mathscr{S}_\lambda^{-1}(M) \in \mathbb{W}^{n \times n}$ be the matrix

$$\mathscr{S}_\lambda^{-1}(M) = (M(\lambda) - \lambda I)^{-1} + \lambda I \in \mathbb{W}^{n \times n}$$

if the inverse $(M(\lambda) - \lambda I)^{-1}$ exists. The matrix $\mathscr{S}_\lambda^{-1}(M)$ is called the *spectral inverse* of the matrix $M(\lambda)$.

We will typically write the spectral inverse of $M \in \mathbb{W}^{n \times n}$ as $\mathscr{S}^{-1}(M)$ unless otherwise noted.

We note that a necessary and sufficient condition for $\mathscr{S}^{-1}(M)$ to exist is that the matrix $M(\lambda) - \lambda I$ be invertible. For instance, the matrix

$$M = \begin{bmatrix} \lambda & 0 \\ 0 & \lambda \end{bmatrix} \in \mathbb{W}^{2 \times 2}$$

cannot be spectrally inverted. However, if M has a spectral inverse, then the following holds.

Theorem 1.4. *Suppose $M(\lambda) \in \mathbb{W}^{n \times n}$ has a spectral inverse $\mathscr{S}^{-1}(M)$. Then*

$$\sigma\left(\mathscr{S}^{-1}(M)\right) = \sigma^{-1}(M) \text{ and } \sigma^{-1}\left(\mathscr{S}^{-1}(M)\right) = \sigma(M).$$

Proof. Let $M(\lambda) \in \mathbb{W}^{n \times n}$ with spectral inverse $\mathscr{S}^{-1}(M)$. Note that

$$\det\left((\mathscr{S}^{-1}(M) - \lambda I)(M - \lambda I)\right) = \det\left((M - \lambda I)^{-1}(M - \lambda I)\right) = \det(I) = 1.$$

Since the determinant is multiplicative, it follows that

$$\det(\mathscr{S}^{-1}(M) - \lambda I) = \det(M - \lambda I)^{-1},$$

and the result follows. □

As noted, a matrix $M \in \mathbb{W}^{n \times n}$ may or may not have a spectral inverse. However, if $M \in \mathbb{W}_\pi^{n \times n}$, then the proof of Theorem 1.2 implies that $M - \lambda I$ is invertible. Therefore, $\mathscr{S}^{-1}(M)$ exists. This result is stated as the following lemma.

Lemma 1.3. *If $M(\lambda) \in \mathbb{W}_\pi^{n \times n}$, then $M(\lambda)$ has a spectral inverse.*

Example 1.7. Let $M \in \mathbb{W}_\pi^{4\times 4}$ be the matrix given by

$$M = \begin{bmatrix} \frac{1}{\lambda} & \frac{1}{\lambda} & 0 & 0 \\ 0 & \frac{1}{\lambda} & 1 & 0 \\ 0 & 0 & \frac{1}{\lambda} & 0 \\ 0 & 0 & 0 & \frac{1}{\lambda} \end{bmatrix},$$

for which

$$\det\left(M(\lambda) - \lambda I\right) = \frac{\lambda^8 - 4\lambda^6 + 6\lambda^4 - 4\lambda^2 + 1}{\lambda^4}.$$

As one can calculate, the spectral inverse $\mathscr{S}^{-1}(M)$ is the matrix

$$\mathscr{S}^{-1}(M) = \begin{bmatrix} \frac{-\lambda}{\lambda^2-1} & \frac{-\lambda}{(\lambda^2-1)^2} & \frac{-\lambda^2}{(\lambda^2-1)^3} & \frac{-\lambda^3}{(\lambda^2-1)^4} \\ 0 & \frac{-\lambda}{\lambda^2-1} & \frac{-\lambda^2}{(\lambda^2-1)^2} & \frac{-\lambda^3}{(\lambda^2-1)^3} \\ 0 & 0 & \frac{-\lambda}{\lambda^2-1} & \frac{-\lambda^2}{(\lambda^2-1)^2} \\ 0 & 0 & 0 & \frac{-\lambda}{\lambda^2-1} \end{bmatrix} + \lambda I.$$

Taking the determinant of $\mathscr{S}^{-1}(M) - \lambda I$, one has

$$\det\left(\mathscr{S}^{-1}(M) - \lambda I\right) = \frac{\lambda^4}{\lambda^8 - 4\lambda^6 + 6\lambda^4 - 4\lambda^2 + 1}.$$

That is, $\det\left(\mathscr{S}^{-1}(M) - \lambda I\right) = \det(M(\lambda) - \lambda I)^{-1}$.

Observe that for every $M \in \mathbb{W}_\pi^{n\times n}$, the spectral inverse $\mathscr{S}^{-1}(M)$ does not belong to $\mathbb{W}_\pi^{n\times n}$. Therefore, we have no guarantee via Theorem 1.2 that $\mathscr{S}^{-1}(M)$ can be isospectrally reduced. As it turns out, though, the following holds.

Theorem 1.5 (Reductions of the Spectral Inverse). *For $M(\lambda) \in \mathbb{W}_\pi^{n\times n}$, suppose that $N \supset S_1 \supset, \ldots, \supset S_m$, where $S_m = S$ is nonempty. Then*

(i) $\mathscr{R}\left(\mathscr{S}^{-1}(M); S\right)$ *exists;*
(ii) $\mathscr{R}(\mathscr{S}^{-1}(M); S_1, \ldots, S_m) = \mathscr{R}(\mathscr{S}^{-1}(M); S_m)$; *and*
(iii) $\mathscr{R}(\mathscr{S}^{-1}(M); S) = (M - \lambda I)^{-1} / \left[(M - \lambda I)^{-1}\right]_{\bar{S}\bar{S}} + \lambda I$.

Proof. For $M \in \mathbb{W}_\pi^{n\times n}$, suppose S and $\bar{S}$ form a nonempty partition of N. By Lemma 1.3, the matrix $\mathscr{S}^{-1}(M)$ exists, and

$$\mathscr{S}^{-1}(M) - \lambda I = (M - \lambda I)^{-1} \in \mathbb{W}_\pi^{n\times n}.$$

By considering the submatrices of the matrix in the previous equation, we find that $[\mathscr{S}^{-1}(M)]_{SS} - \lambda I$, $[\mathscr{S}^{-1}(M)]_{S\bar{S}}$, $[\mathscr{S}^{-1}(M)]_{\bar{S}S}$, and $[\mathscr{S}^{-1}(M)]_{\bar{S}\bar{S}} - \lambda I$ all have

1.4 Spectral Inverse

entries in $\mathbb{W}_\pi$. Moreover, $[\mathscr{S}^{-1}(M)]_{\bar{S}\bar{S}} - \lambda I$ is not identically zero, so its inverse exists. We deduce that the reduction of $\mathscr{S}^{-1}(M)$ over S exists and is given by

$$\mathscr{R}(\mathscr{S}^{-1}(M); S) - \lambda I$$
$$= ([\mathscr{S}^{-1}(M)]_{SS} - \lambda I) - [\mathscr{S}^{-1}(M)]_{S\bar{S}} \left([\mathscr{S}^{-1}(M)]_{\bar{S}\bar{S}} - \lambda I\right)^{-1} [\mathscr{S}^{-1}(M)]_{\bar{S}S}.$$

Moreover, $\mathscr{R}(M; S) \in \mathbb{W}_\pi^{|S| \times |S|}$.

To prove (iii), notice that we have

$$[\mathscr{S}^{-1}(M)]_{SS} - \lambda I = \left[(M - \lambda I)^{-1}\right]_{SS}$$
$$[\mathscr{S}^{-1}(M)]_{\bar{S}S} = [\mathscr{S}^{-1}(M) - \lambda I]_{\bar{S}S} = \left[(M - \lambda I)^{-1}\right]_{\bar{S}S}$$
$$[\mathscr{S}^{-1}(M)]_{S\bar{S}} = \left[(M - \lambda I)^{-1}\right]_{S\bar{S}} \text{ and}$$
$$[\mathscr{S}^{-1}(M)]_{\bar{S}\bar{S}} - \lambda I = \left[(M - \lambda I)^{-1}\right]_{\bar{S}\bar{S}}.$$

These relations imply (iii).

Substituting each submatrix M_{RC} in the proof of Lemma 1.2 by the matrix

$$\mathscr{S}^{-1}(M)_{RC} = \begin{cases} (M - \lambda I)_{RC}^{-1} + \lambda I & \text{if } R = C, \\ (M - \lambda I)_{RC}^{-1} & \text{otherwise}, \end{cases}$$

and then following the proof of Theorem 1.3 using $\mathscr{S}^{-1}(M)$ instead of M yields a proof of part (ii). □

In summary, if $M \in \mathbb{W}_\pi^{n \times n}$, then it is possible to reduce both M and its spectral inverse $\mathscr{S}^{-1}(M)$ over every nonempty index set $S \subset N$. Moreover, the eigenvalues and inverse eigenvalues of these reduced matrices can be found using Theorem 2.1.

Chapter 2
Dynamical Networks and Isospectral Graph Reductions

This is a fundamental chapter of the book. It deals with networks, which are here considered as graphs, and is built on the theory developed in the previous chapter, on matrices.

Although, the dynamical networks described in Chap. 3 are richer objects than their weighted adjacency matrices, the latter still carry the most important information about a dynamical network. Indeed, from a theoretical point of view, a network's weighted adjacency matrix describes a linearization of the network's dynamics, which in applications is often the only network information available. In fact, it is not uncommon to have only the unweighted adjacency matrix of a network.

In this chapter, we introduce the notion of an isospectral graph reduction, which is based on the essential idea of the branch structure of a graph (network). A graph's collection of branches, which are related to the more familiar concepts of paths and cycles, form the foundation of the isospectral transformations considered in this book, both in this chapter and later in applications of this theory.

The isospectral graph reductions that we study in this chapter allow us to reduce the size of a graph (number of vertices) while maintaining the graph's spectrum, up to a known set. Besides these reductions, we also introduce and analyze a number of other graph transformations that affect the graph's spectrum in a specific way. These include the operations of branch expansions, reweightings, and mergings.

These graph transformations can be used to simplify the structure of a graph (network) while preserving the graph's (network's) spectrum, up to a known set, as well as its collection of edge weights. For example, it is possible to reduce a network in which all edges have weight 1 to a smaller network with the same property. That is, these isospectral transformations do not somehow shift the complexity of the original network to the edge weights of the reduced network.

We begin this chapter by introducing the fundamental operation of isospectral graph reduction. After analyzing its properties, we show that a sequence of isospectral reductions of the same graph results in the same reduced network, regardless of the order in which the vertices were removed, i.e., the result depends only on the final collection of vertices.

It is worthwhile mentioning that the procedure of isospectral graph reduction can be applied not only to the vertices of graphs (elements of networks) but also to the graph's edges. To do this, one need only consider the line (dual) graph, in which all edges of the initial graph are vertices, and apply the same procedure to it. Because all the results and operations are the same, we will deal only with removal of vertices.

We also introduce a new equivalence relation on the collection of graphs (networks) that we consider. Namely, two graphs (networks) are said to be *spectrally equivalent* if they can be isospectrally reduced to one and the same graph (network). The basic idea here is that two graphs may look very different but be spectrally equivalent, suggesting that the corresponding networks have similar dynamics.

Although each of these results is rigorously proved, we have attempted to make the exposition as visual and accessible as possible. With this goal in mind, the definitions and results of this chapter are illustrated by numerous examples and figures.

2.1 Dynamical Networks as Graphs

To begin, we note that to each dynamical network there is an associated weighted directed graph G, which we call the network's *graph of interactions*. As the name suggests, this graph describes the various interactions among the network's elements. In network theory, the unweighted version of this graph is often called the network's *topology*. Here, we do not use this term, since it might be confused with the unrelated branch of mathematics called topology.

The graph $G = (V, E, \omega)$ is composed of a *vertex set* V, a set of directed *edges* E, and a function ω that gives a weight to each edge in E. The *vertex set* V represents the elements of the network, and the *edge set* E, the interactions among these elements. Because it is assumed that the graph G corresponds to a network, we consider only *finite graphs*, or those graphs in which V and E are finite and nonempty.

For $V = \{v_1, \ldots, v_n\}$, we let e_{ij} denote the edge from vertex v_i to v_j. The edge e_{ij} is an element of E if the ith network element interacts with (or directly influences) the jth network element. The function ω gives the edge weights of G, where $\omega(e_{ij})$, or the edge weight of e_{ij}, corresponds to the strength of the interaction between the ith and jth elements of a network. We adopt the standard convention that each edge of G has a nonzero weight. More formally, $\omega(e_{ij}) = 0$ if and only if $e_{ij} \notin E$, or the ith element of the network does not directly influence the jth network element.

Suppose the graph $G = (V, E, \omega)$ has the vertex set $V = \{v_1, \ldots, v_n\}$. We define the $n \times n$ matrix $M(G)$ by

$$M(G)_{ij} = \omega(e_{ij}).$$

2.1 Dynamical Networks as Graphs

The matrix $M(G)$ is called the *weighted adjacency matrix* of G. If G is the graph of interactions of a dynamical network, we say that the eigenvalues of the matrix $M(G)$ make up the spectrum of this dynamical network. In later chapters, we will connect the spectrum of a network with its dynamics. For now, we simply assume that to each dynamical network there is an associated graph G with adjacency matrix $M(G)$.

To compute the spectrum of a network, we need to know the weights of its graph of interactions $G = (V, E, \omega)$. At this point, we have yet to formally define the *weight set* $\omega(E) = \{\omega(e) : e \in E\}$ that will be considered here. In practice, one often chooses $\omega(E)$ to be some subset of the real numbers. For example, if we consider the graph G to be an unweighted graph, then $M(G)$ is the matrix with 0–1 entries given by

$$M(G)_{ij} = \begin{cases} 1 & \text{if } e_{ij} \in E, \\ 0 & \text{otherwise.} \end{cases}$$

However, the set of weights we will use is not a subset of the real or even complex numbers but the set of rational functions $\mathbb{W}$, defined in Chap. 1. The class of graphs we consider is defined as follows.

Definition 2.1. Let $\mathbb{G}$ be the collection of weighted directed graphs given by

$$\mathbb{G} = \{G = (V, E, \omega) : \omega : E \to \mathbb{W}\}.$$

The *spectrum*, or set of eigenvalues, of a graph $G \in \mathbb{G}$ is the set $\sigma(G) = \sigma(M(G))$. The *inverse spectrum* of G is the set $\sigma^{-1}(G) = \sigma^{-1}(M(G))$.

Equivalently, one could define $\mathbb{G}$ as the set of all graphs $G = (V, E, \omega)$ for which $M(G) \in \mathbb{W}^{n \times n}$ for some $n \in \mathbb{N}$. In either case, Definition 1.1 gives the spectrum and inverse spectrum of the graph $G \in \mathbb{G}$.

To stress the generality of considering the set $\mathbb{G}$, we note that graphs that are either undirected or have parallel edges can be considered graphs in $\mathbb{G}$. In particular, an undirected graph can be made into a directed graph by orienting each of its edges in both directions. If a graph G has multiple edges between two vertices, a single edge can be put in their place whose weight is the sum of the weights of the edges that it is replacing.

Example 2.1. Consider the graphs G and H shown in Fig. 2.1. The weighted adjacency matrices $M(G) \in \mathbb{W}^{4 \times 4}$ and $M(H) \in \mathbb{C}^{6 \times 6}$ are given in (1.4) and (1.5), respectively. Hence,

$$\sigma(G) = \sigma(H) = \{-2, -2, -1 \pm i, -1 \pm \sqrt{3}\}.$$

That is, G and H have the same spectrum, although H has more vertices than G.

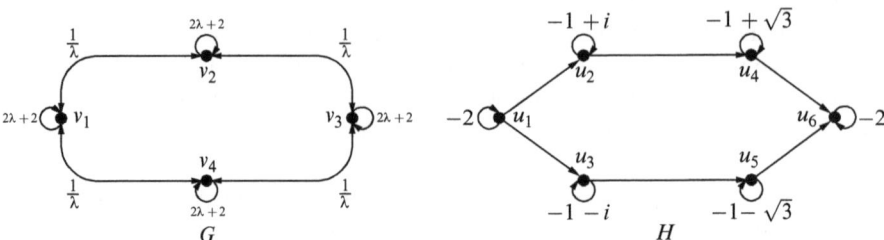

Fig. 2.1 The graph $G \in \mathbb{W}^{4\times 4}$ (*left*) given in Example 2.1 and $H \in \mathbb{C}^{6\times 6}$, where $\sigma(G) = \sigma(H)$

Again, our main question is whether a graph can be reduced to a smaller graph in such a way that its spectrum is preserved. Equivalently, we ask whether there is a way to reduce a graph H to a graph G such that $M(H)$ and $M(G)$ have the same eigenvalues.

The answer from Chap. 1 is that this is indeed possible. The goal, then, of this chapter is to describe what isospectral graph reductions means in terms of the original graph (network) and second how to use this understanding to develop other isospectral transformations of a graph (network).

One point we hope to make is that there are a number of useful isospectral transformations that can be developed using the theory of isospectral graph reductions as an initial starting point. In this chapter, we introduce a handful of such transformations that will be useful later for different purposes. The major point is that one can develop the most relevant transformation for a given problem with respect to both analytic simplicity and effectiveness as a computational tool.

2.2 Isospectral Graph Reductions

Because of the structural complexity and considerable size of many networks, the corresponding graph of interactions $G = (V, E, \omega)$ may have an extremely large vertex set V as well as a large and irregular set of edges. To reduce this complexity while maintaining the network's spectral properties, we introduce the concept of isospectral graph reductions, which is related to isospectral matrix reduction, studied in the previous chapter.

The spectrum of a graph is intimately related to its structure. Specifically, knowing the graph's path and cycle structure along with its weights gives us enough information to compute the graph's spectrum. Simply put, a path is a sequence of distinct vertices that can be traversed by moving along edges of the graph. A cycle is a path that begins and ends at the same vertex.

More formally, a *path* P in the graph $G = (V, E, \omega)$ is an ordered sequence of distinct vertices $P = v_1, \ldots, v_m \in V$ such that $e_{i,i+1} \in E$ for $1 \leq i \leq m - 1$. We call the vertices $v_2, \ldots, v_{m-1}$ of P the *interior vertices* of P. If the vertices v_1

2.2 Isospectral Graph Reductions

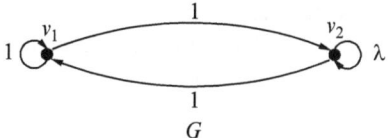

Fig. 2.2 The graph $G \in \mathbb{W}^{4 \times 4}$ with adjacency matrix $M(G)$ considered in Example 1.5

and v_m are the same, then P is a *cycle*. If it is the case that a cycle contains a single vertex, then we call this cycle a *loop*. In addition, since v_i is a loop of G if and only if $e_{ii} \in E$, we may refer to the edge e_{ii} as a loop.

The main idea behind the isospectral reduction of a graph $G = (V, E, \omega)$ is that we reduce G to a smaller graph on some subset $S \subset V$. Equating a graph G with its adjacency matrix $M(G)$, we note that formally, an isospectral reduction of G could be defined as the graph with the "reduced" adjacency matrix $\mathscr{R}(M(G); S)$. Indeed, this will be the case. However, to investigate how the structure of a graph is affected by an isospectral reduction, we deliberately limit the type of vertex sets over which we can reduce a graph. Such sets, called structural sets, are defined as follows.

Definition 2.2. Let $G = (V, E, \omega) \in \mathbb{G}$. A nonempty vertex set $S \subset V$ is a *structural set* of G if

(i) each cycle of G that is not a loop contains a vertex in S; and
(ii) $\omega(e_{ii}) \neq \lambda$ for each $v_i \in \bar{S} = V - S$.

For $G = (V, E, \omega)$, suppose S is a subset of the vertex set V. Then the graph $G|_S$ is called the *subgraph* of G *induced* over the vertex set S and is given by

$$G|_S = (S, \mathscr{E}, \mu) \text{ where } \mathscr{E} = \{e_{ij} \in E : v_i, v_j \in S\} \text{ and } \mu = \omega|_\mathscr{E}.$$

If S is a structural set of G, part (i) of Definition 2.2 says that the subgraph $G|_{\bar{S}}$ has no cycles except possibly for loops. Part (ii) of Definition 2.2 is the formal assumption that the loops of the vertices in $\bar{S}$, i.e., the complement of S, do not have weight equal to $\lambda \in \mathbb{W}$. That is, these loops are not weighted by the rational function $\lambda/1 \in \mathbb{W}$.

Consider the graph $G \in \mathbb{G}$ with adjacency matrix

$$M(G) = \begin{bmatrix} 1 & 1 \\ 1 & \lambda \end{bmatrix},$$

as in Example (1.5). The graph G is shown in Fig. 2.2. Notice that $\omega(e_{22}) = \lambda$. Hence, the set $S = \{v_2\}$ is not a structural set of G.

For $G \in \mathbb{G}$, we let $st(G)$ denote the set of all structural sets of the graph G. The idea behind the notion of a structural set $S \in st(G)$ is the following. Every random walk along edges of G that begins at a vertex in S eventually finds its way to another vertex of S, if we ignore loops. Therefore, a structural set allows us essentially to partition a random walk on G into finite paths and cycles that begin and end with vertices of S. We give these paths and cycles the following name.

Definition 2.3. Suppose $G = (V, E, \omega)$ with the structural set $S = \{v_1, \ldots, v_m\}$. Let $\mathscr{B}_{ij}(G; S)$ be the set of paths from v_i to v_j, or cycles if $i = j$, with no interior vertices in S. We call a path or cycle $\beta \in \mathscr{B}_{ij}(G; S)$ a *branch* of G with respect to S. We let

$$\mathscr{B}_S(G) = \bigcup_{1 \le i,j \le m} \mathscr{B}_{ij}(G; S)$$

denote the *branch set* of all branches of G with respect to S.

If $\beta = v_1, \ldots, v_m$ is a branch of G with respect to S and $m > 2$, we define

$$\mathscr{P}_\omega(\beta) = \omega(e_{12}) \prod_{i=2}^{m-1} \frac{\omega(e_{i,i+1})}{\lambda - \omega(e_{ii})}. \tag{2.1}$$

For $m = 1, 2$, we let $\mathscr{P}_\omega(\beta) = \omega(e_{1m})$. We call $\mathscr{P}_\omega(\beta)$ the *branch product* of β.

Notice that assumption (ii) in Definition 2.2 implies that the branch product of every $\beta \in \mathscr{B}_S(G)$ is always defined and is a rational function in $\mathbb{W}$. In fact, the reason we require that part (ii) of Definition 2.2 hold is to ensure that the branch product of each branch in $\mathscr{B}_S(G)$ exists.

To *isospectrally reduce* a graph over the set $S \in st(G)$, we replace each branch $\mathscr{B}_{ij}(G; S)$ with a single edge $e_{ij} \in \mathscr{E}$. The following definition specifies the weights of these edges.

Definition 2.4. Let $G = (V, E, \omega)$ with structural set $S = \{v_1, \ldots, v_m\}$. Define the edge weights

$$\mu(e_{ij}) = \begin{cases} \sum_{\beta \in \mathscr{B}_{ij}(G;S)} \mathscr{P}_\omega(\beta) & \text{if } \mathscr{B}_{ij}(G; S) \ne \emptyset, \\ 0 & \text{otherwise,} \end{cases} \quad \text{for } 1 \le i, j \le m. \tag{2.2}$$

The graph $\mathscr{R}_S(G) = (S, \mathscr{E}, \mu)$ in which $e_{ij} \in \mathscr{E}$ if $\mu(e_{ij}) \ne 0$ is the *isospectral reduction* of G over S.

Observe that $\mu(e_{ij})$ in Definition 2.4 is the weight of the edge e_{ij} in $\mathscr{R}_S(G)$. Moreover, since $\mathbb{W}$ is closed under both addition and multiplication, it follows that the edge weights $\mu(e_{ij})$ of $\mathscr{R}_S(G)$ are also in the set $\mathbb{W}$. Hence the isospectral reduction $\mathscr{R}_S(G)$ is again a graph in $\mathbb{G}$.

Example 2.2. Consider the graph $G = (V, E, \omega)$ given in Fig. 2.3 (left), where G is an unweighted graph, i.e., each edge of G is given unit weight. Note that the vertex set $S = \{v_1, v_3\} \subset V$ is a structural set of G, since

(i) the three nonloop cycles of G, namely v_1, v_2, v_3, v_4, v_1; v_1, v_5, v_1; and v_3, v_6, v_3, each contains a vertex in S; and

2.2 Isospectral Graph Reductions

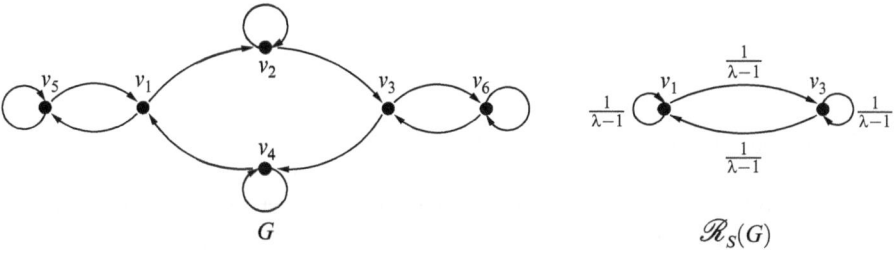

Fig. 2.3 Reduction of G over $S = \{v_1, v_3\}$ from Example 2.2, where each edge in G has unit weight

(ii) the loop weights of vertices in $\bar{S} = \{v_2, v_4, v_5, v_6\}$ are $\omega(e_{22}) = 1$, $\omega(e_{44}) = 1$, $\omega(e_{55}) = 1$, and $\omega(e_{66}) = 1$ respectively.

Hence, $\omega(e_{ii}) = 1 \in \mathbb{W}$ is not equal to the rational function $\lambda/1 \in \mathbb{W}$ for each $v_i \in \bar{S}$.

In contrast, the vertex set $T = \{v_1, v_2, v_5\}$ is not a structural set of G, since the (nonloop) cycle v_3, v_6, v_3 does not contain a vertex of T. Phrased another way, the random walk $v_2, v_3, v_6, v_3, v_6, \ldots$ cannot be partitioned into finite paths and cycles that begin and end with vertices in T.

Returning to the structural set S, we see that the branches in $\mathscr{B}_S(G)$ are respectively given by $\mathscr{B}_{11}(G; S) = \{v_1, v_5, v_1\}$, $\mathscr{B}_{13}(G; S) = \{v_1, v_2, v_3\}$, $\mathscr{B}_{31}(G; S) = \{v_3, v_4, v_1\}$, and $\mathscr{B}_{33}(G; S) = \{v_3, v_6, v_3\}$. Using (2.1), we conclude that the branch product of each branch is given by

$$\mathscr{P}_\omega(v_1, v_5, v_1) = \mathscr{P}_\omega(v_1, v_2, v_3) = \mathscr{P}_\omega(v_3, v_4, v_1) = \mathscr{P}_\omega(v_3, v_6, v_3) = \frac{1}{\lambda - 1}.$$

Equation (2.2) then gives each edge of $\mathscr{R}_S(G) = (S, \mathscr{E}, \mu)$ the weight

$$\mu(e_{11}) = \mu(e_{13}) = \mu(e_{31}) = \mu(e_{33}) = \frac{1}{\lambda - 1}.$$

Since each edge weight is nonzero, the edge set $\mathscr{E}$ of $\mathscr{R}_S(G)$ is $\mathscr{E} = \{e_{11}, e_{13}, e_{31}, e_{33}\}$. In particular, an edge of $\mathscr{E}$ need not be an edge of E. The graph $\mathscr{R}_S(G)$ is shown in Fig. 2.3 (right).

To relate an isospectral graph reduction to an isospectral matrix reduction, we need a way of connecting a structural set with a submatrix. For the graph $G = (V, E, \omega)$, suppose that $S, T \subset V$ are nonempty. By a slight abuse of notation, let $M(G)_{ST}$ denote the submatrix of the graph's weighted adjacency matrix with rows indexed by the vertices in S and columns indexed by the vertices in T. This allows us to state the following fundamental relation between isospectral graph and matrix reductions.

Theorem 2.1. *Let S be a structural set of the graph $G \in \mathbb{G}$. Then*

$$M(\mathscr{R}_S(G)) = \mathscr{R}(M(G); S).$$

Before proving Theorem 2.1, we note the following. A matrix $P \in \mathbb{C}^{n \times n}$ is a *permutation matrix* if each row and column of P has exactly one entry that is 1 with all other entries equal to 0. We will use the fact that every permutation matrix is invertible. Moreover, for $G = (V, E, \omega) \in \mathbb{G}$, the matrix $PM(G)P^{-1}$ is the adjacency matrix of G in which the vertices V have been relabeled by the permutation P.

Proof. Let $G = (V, E, \omega)$, where $V = \{v_1, \ldots, v_n\}$. Without loss in generality, suppose $S = \{v_1, \ldots, v_m\}$ is a structural set of the graph G. Since the subgraph $G|_{\bar{S}}$ has no cycles except loops, the vertices of $G|_{\bar{S}}$ can be relabeled such that the matrix $M(G|_{\bar{S}})$ is upper triangular. By assumption (ii) of Definition 2.2, the diagonal entries $M(G|_{\bar{S}})_{ii}$ are not equal to λ, implying

$$(M(G|_{\bar{S}}) - \lambda I)_{ii} \neq 0 \text{ for all } m < i \leq n.$$

Hence, $M(G)_{\bar{S}\bar{S}} - \lambda I$ is an upper triangular matrix with nonzero diagonal and is therefore invertible.

Letting $M = M(G)$, the upper triangular matrix $M_{\bar{S}\bar{S}} - \lambda I$ can be written as

$$M_{\bar{S}\bar{S}} - \lambda I = D(I + N),$$

where $D = \text{diag}[M_{m+1,m+1} - \lambda \ldots M_{nn} - \lambda]$ and $N \in \mathbb{W}^{(n-m) \times (n-m)}$ is the nilpotent matrix given by

$$N_{ij} = \begin{cases} \frac{M_{i+m,j+m}}{M_{i+m,i+m}-\lambda} & \text{for } i < j, \\ 0 & \text{otherwise.} \end{cases} \tag{2.3}$$

The inverse of $M_{\bar{S}\bar{S}} - \lambda I$ is then

$$(I + N)^{-1} D^{-1} = (I + N - N^2 - \cdots + (-1)^{n-m-1} N^{n-m-1}) D^{-1}, \tag{2.4}$$

where $D^{-1} = \text{diag}[\frac{1}{M_{m+1,m+1}-\lambda} \cdots \frac{1}{M_{nn}-\lambda}]$.

For $1 \leq \ell \leq n - m - 1$, the matrix N^ℓ is given by

$$N^\ell_{ij} = \begin{cases} \sum \prod_{k=1}^{\ell} N_{i_{k-1}, i_k} & \text{for } i < j, \\ 0 & \text{otherwise,} \end{cases}$$

where the sum is taken over all strictly increasing $(\ell + 1)$-tuples $i = i_0 < \cdots < i_\ell = j$ starting at i and ending at j. Hence for $1 \leq i, j \leq m$, we have

2.2 Isospectral Graph Reductions

$$M_{ij} + (M_{S\bar{S}}(M_{\bar{S}\bar{S}} - \lambda I)^{-1} M_{\bar{S}S})_{ij}$$
$$= M_{ij} + (M_{S\bar{S}}(I + N - N^2 - \cdots + (-1)^{n-m-1} N^{n-m-1}) D^{-1} M_{\bar{S}S})_{ij}$$
$$= M_{ij} + \sum_{p=m+1}^{n} \sum_{q=m+1}^{n} M_{pi} \left(\sum_{\ell=0}^{n-m-1} (-1)^\ell N^\ell D^{-1} \right)_{pq} M_{qj}.$$

Observe that the entries of the matrix $(-1)^\ell N^\ell D^{-1}$ are

$$(-1)^\ell (N^\ell D^{-1})_{ij} = \begin{cases} \left(\sum \prod_{k=1}^{\ell} \frac{M_{i_{k-1}+m, i_k + m}}{\lambda - M_{i_{k-1}+m, i_{k-1}+m}} \right) \left(\frac{1}{\lambda - M_{j+m, j+m}} \right) & \text{for } i < j, \\ 0 & \text{otherwise}, \end{cases}$$

where as before, the sum is taken over all increasing $(\ell + 1)$-tuples $i = i_0 < \cdots < i_\ell = j$ starting at i and ending at j. Therefore,

$$M_{ij} + (M_{S\bar{S}}(M_{\bar{S}\bar{S}} - \lambda I)^{-1} M_{\bar{S}S})_{ij} = M_{ij} + \sum \left(\sum_{\ell=0}^{n-m-1} \frac{M_{ip}}{\lambda - M_{pp}} \prod_{k=1}^{\ell} \frac{M_{i_{k-1}, i_k}}{\lambda - M_{i_k, i_k}} M_{qj} \right),$$

where the sum is taken over all increasing $(\ell+1)$-tuples $p = i_0 < i_1 < \cdots < i_\ell = q$ for $m < p, q \le n$.

Under the assumption that $M(G)_{\bar{S}\bar{S}} - \lambda I$ is upper triangular, the graph G has the following property: if $\beta = v_i, v_{i_2}, \ldots, v_{i_{\ell-1}}, v_j \in \mathcal{B}_S(G)$, then $i_2 < \cdots < i_{\ell-1}$, where $i, j \le m$ and $i_2, \ldots, i_{\ell-1} > m$.

Thus, the branch product of β can be written as

$$\mathcal{P}_\omega(\beta) = \frac{M_{i, i_2}}{\lambda - M_{i_2, i_2}} \prod_{k=1}^{\ell} \frac{M_{i_{k-1}, i_k}}{\lambda - M_{i_k, i_k}} M_{i_{\ell-1}, j}.$$

Summing over all branches of $\mathcal{B}_S(G)$ from v_i to v_j, we arrive at

$$M(\mathcal{R}_S(G))_{ij} = M_{ij} + \sum \left(\sum_{\ell=0}^{n-m-1} \frac{M_{ip}}{\lambda - M_{pp}} \prod_{k=1}^{\ell} \frac{M_{i_{k-1}, i_k}}{\lambda - M_{i_k, i_k}} M_{qj} \right).$$

Therefore, the weighted adjacency matrix of $\mathcal{R}_S(G)$ is

$$M(\mathcal{R}_S(G)) = M(G)_{SS} - M(G)_{S\bar{S}} (M(G)_{\bar{S}\bar{S}} - \lambda I)^{-1} M(G)_{\bar{S}S} = \mathcal{R}(M(G), S),$$

which completes the proof. □

The reason we consider both graph and matrix reductions is that both points of view have their advantages. For one, graphs allow one to analyze network structure visually, while matrices allow for easy storage and manipulation of this network information. More to the point, equation (2.2) indicate how the spectrum of a graph is preserved in terms of the graph's branch structure. This observation will allow us, in Sects. 2.5 and 2.6 of this chapter, to develop other types of isospectral transformations. These transformations will be based on manipulating a graph's path and cycle structure in ways that modify the graph's spectrum in a predictable way.

For now, we note that if S is a structural set of the graph $G \in \mathbb{G}$, then the isospectral reduction $\mathcal{R}_S(G)$ is also a graph in $\mathbb{G}$. Hence, both G and $\mathcal{R}_S(G)$ have well-defined spectra. The relation between $\sigma(G)$ and $\sigma(\mathcal{R}_S(G))$ is given in the following corollary of Theorems 1.1 and 2.1.

Corollary 2.1. *Let S be a structural set of the graph $G \in \mathbb{G}$. Then*

$$\sigma(\mathcal{R}_S(G)) = \left(\sigma(G) \cup \sigma^{-1}(G|_{\bar{S}})\right) - \left(\sigma(G|_{\bar{S}}) \cup \sigma^{-1}(G)\right); \text{ and}$$

$$\sigma^{-1}(\mathcal{R}_S(G)) = \left(\sigma(G|_{\bar{S}}) \cup \sigma^{-1}(G)\right) - \left(\sigma(G) \cup \sigma^{-1}(G|_{\bar{S}})\right).$$

Suppose that $G = (V, E, \omega)$ has the structural set S. Because the graph $G|_{\bar{S}}$ has a particular form, it is possible to compute both $\sigma(G|_{\bar{S}})$ and $\sigma^{-1}(G|_{\bar{S}})$ quickly. That is, since $M(G|_{\bar{S}}) - \lambda I$ is similar to an upper triangular matrix via some permutation, it follows that

$$\det\left(M(G|_{\bar{S}}) - \lambda I\right) = \prod_{v_i \in \bar{S}} (\omega(e_{ii}) - \lambda). \tag{2.5}$$

Since the product $\prod_{v_i \in \bar{S}} (\omega(e_{ii}) - \lambda) = p(\lambda)/q(\lambda)$ for some $p(\lambda)/q(\lambda) \in \mathbb{W}$ then

$$\sigma(G|_S) = \{\lambda \in \mathbb{C} : p(\lambda) = 0\} \text{ and } \sigma^{-1}(G|_S) = \{\lambda \in \mathbb{C} : q(\lambda) = 0\}. \tag{2.6}$$

For complex-valued matrices, Corollary 2.1 together with (2.5) and (2.6) implies the following corollary.

Corollary 2.2. *Let S be a structural set of the graph $G \in \mathbb{G}$. If $M(G) \in \mathbb{C}^{n \times n}$, then*

(i) $\sigma^{-1}(G) = \emptyset$ and $\sigma^{-1}(G|_{\bar{S}}) = \emptyset$;
(ii) $\sigma(G|_{\bar{S}}) = \{\omega(e_{ii}) : v_i \in \bar{S}\}$;
(iii) $\sigma(\mathcal{R}_S(G)) = \sigma(G) - \sigma(G|_{\bar{S}})$; and
(iv) $\sigma^{-1}(\mathcal{R}_S(G)) = \sigma(G|_{\bar{S}}) - \sigma(G)$.

In many applications, the graphs (matrices) that are used have real or positive weights (entries). If $G = (V, E, \omega)$ has complex-valued weights and $S \in st(G)$, then part (iii) of Corollary 2.2 states that the spectra of $\mathcal{R}_S(G)$ and G differ at most by the spectrum of $G|_{\bar{S}}$. Moreover, part (ii) of Corollary 2.2 states that the spectrum $\sigma(G|_{\bar{S}})$ consists of the weights of the loops e_{ii} for $v_i \in \bar{S}$ and is therefore easily identified.

2.2 Isospectral Graph Reductions

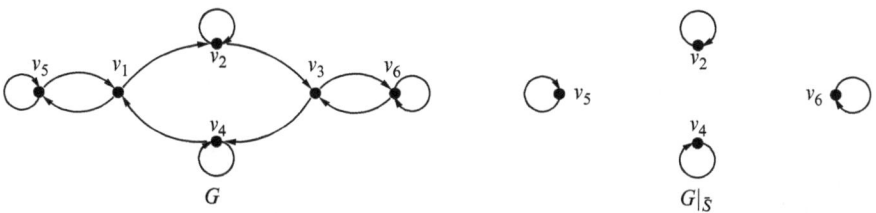

Fig. 2.4 The restriction of the graph G in Example 2.5 to $\bar{S} = \{v_2, v_4, v_5, v_6\}$, where each edge in each of G and $G|_{\bar{S}}$ has unit weight

Similar to Theorem 1.1 for matrices, Corollary 2.1 describes exactly which eigenvalues we may gain from an isospectral reduction and which me may lose. In this way, an isospectral reduction of a graph preserves the spectral information of the original graph. This will be important, for instance, in Sect. 2.6, where we consider isospectral reductions that do not affect the nonzero eigenvalues of a graph.

Example 2.3. Let G be the graph considered in Example 2.2. As previously shown, the vertex set $S = \{v_1, v_3\}$ is a structural set of G. Moreover, $M(G) \in \mathbb{C}^{6\times 6}$. Hence, Corollary 2.2 allows us to compute the eigenvalues of the reduced graph $\mathscr{R}_S(G)$ quickly once the eigenvalues of G are known.

As one can compute, the eigenvalues of the graph G are $\sigma(G) = \{2, -1, 1, 1, 1, 0\}$. The restricted graph $G|_{\bar{S}}$, shown in Fig. 2.4 (right), has loop weights $\omega(e_{22}) = 1$, $\omega(e_{44}) = 1$, $\omega(e_{55}) = 1$, and $\omega(e_{66}) = 1$. Corollary 2.2 then implies that $\sigma(G|_{\bar{S}}) = \{1, 1, 1, 1\}$. Additionally, since $\sigma(\mathscr{R}_S(G)) = \sigma(G) - \sigma(G|_{\bar{S}})$, the spectrum of the reduced graph is $\sigma(\mathscr{R}_S(G)) = \{2, -1, 0\}$.

Since the graph $\mathscr{R}_S(G)$ has two vertices, the matrix $M(\mathscr{R}_S(G))$ belongs to $\mathbb{W}_\pi^{2\times 2}$. However, notice that

$$\det\left(M(\mathscr{R}_S(G)) - \lambda I\right) = \frac{\lambda^3 - \lambda^2 - 2\lambda}{\lambda - 1},$$

which is zero for $\lambda = 2, -1, 0$. Similar to Example 1.1, this is an explicit demonstration of the fact that an $n \times n$ matrix in $\mathbb{W}_\pi^{n\times n}$ may have more than n eigenvalues.

Therefore, the effect of reducing G over S is that we lose the eigenvalues $\{1, 1, 1, 1\}$. However, even if $\sigma(G)$ is unknown, we still know the following. The set of eigenvalues $\sigma(G|_{\bar{S}}) = \{1, 1, 1, 1\}$ is the most by which $\sigma(\mathscr{R}_S(G))$ and $\sigma(G)$ can differ.

We note that this example is equivalent to Example 1.3 in Chap. 1. The difference is that this example is taken from the point of view of a graph reduction. This is done to emphasize the differences and similarities between isospectral reductions of matrices and graphs.

We note that for every $G \in \mathbb{G}$, both $\sigma(G|_{\bar{S}})$ and $\sigma^{-1}(G|_{\bar{S}})$ are easily calculated via equation (2.5). Therefore, Corollary 2.1 offers a quick way of computing the eigenvalues of a reduced graph if the spectrum of the original unreduced graph is known.

Before moving on to the following section, on sequential graph reductions, we note that undirected graphs are often studied in the theory of networks. This is the case whenever the interaction between two network elements has the same effect on both of them, i.e., when the interaction between elements is *symmetric*.

Undirected graphs are particular types of directed graphs and as such can be reduced using Definition 2.4. For the moment, we consider the theory of isospectral graph reductions restricted to the class of undirected graphs.

Definition 2.5. A graph $G = (V, E, \omega)$ in $\mathbb{G}$ is an *undirected graph* if whenever the edge e_{ij} is in E, then $e_{ji} \in E$ and $\omega(e_{ij}) = \omega(e_{ji})$.

A consequence of Definition 2.5 is that a graph $G \in \mathbb{G}$ is undirected if and only if its adjacency matrix $M(G)$ is symmetric. Typically, an undirected graph is assumed to have no loops. The reason we allow undirected graphs to be graphs possibly with loops is that it allows us to state the following results more concisely.

Theorem 2.2. *Suppose S is a structural set of the undirected graph $G \in \mathbb{G}$. Then the reduced graph $\mathcal{R}_S(G)$ is an undirected graph.*

Proof. If $G \in \mathbb{G}$ is undirected, then its adjacency matrix $M = M(G)$ is symmetric. For $S \in st(G)$, we have

$$\left(M_{S\bar{S}}(M - \lambda I)_{\bar{S}\bar{S}}^{-1} M_{\bar{S}S}\right)^T = M_{\bar{S}S}^T \left((M - \lambda I)_{\bar{S}\bar{S}}^{-1}\right)^T M_{S\bar{S}}^T = M_{S\bar{S}}(M - \lambda I)_{\bar{S}\bar{S}}^{-1} M_{\bar{S}S},$$

where the second inequality follows from the fact that $M_{\bar{S}S}^T = M_{S\bar{S}}$, $M_{S\bar{S}}^T = M_{\bar{S}S}$, and the fact that the inverse of the symmetric matrix $(M - \lambda I)_{\bar{S}\bar{S}}$ is also symmetric. Hence, the matrix $M_{S\bar{S}}(M - \lambda I)_{\bar{S}\bar{S}}^{-1} M_{\bar{S}S}$ is symmetric.

It then follows that the reduced matrix

$$\mathcal{R}(M; S) = (M - \lambda I)_{SS} - M_{S\bar{S}}(M - \lambda I)_{\bar{S}\bar{S}}^{-1} M_{\bar{S}S}$$

is symmetric, implying that $\mathcal{R}_S(G)$ is an undirected graph. $\square$

Example 2.4. Let G be the undirected graph shown in Fig. 2.5 (left). The edges found in this figure are undirected, but each can be thought of as being two directed edges oriented in opposite directions with equal (unit) weights.

As one can check, the set $S = \{v_1, v_3, v_4, v_6\}$ is a structural set of G. Theorem 2.2 then implies that the reduced graph $\mathcal{R}_S(G)$ is undirected, which can be seen in Fig. 2.5 (right).

Given a graph $G = (V, E, \omega)$, recall that two vertices $v_i, v_j \in V$ are *adjacent* if $e_{ij} \in E$. Since two adjacent vertices form a cycle in an undirected graph, this

2.3 Sequential Graph Reductions

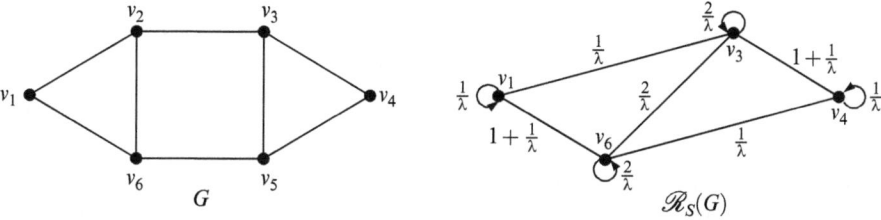

Fig. 2.5 The undirected graph G with unit edge weights and its reduction $\mathcal{R}_S(G)$, where the structural set S is given by $S = \{v_1, v_3, v_4, v_6\}$

naturally restricts which sets can be its structural sets. In fact, this observation allows us to characterize which sets can and cannot be structural sets of an undirected graph.

Theorem 2.3. *Let $G = (V, E, \omega)$ be an undirected graph. Then $S \in st(G)$ if and only if no two vertices $v_i, v_j \in \bar{S}$ are adjacent and $\omega(e_{kk}) \neq \lambda$ for all $v_k \in \bar{S}$.*

Proof. For $S \in st(G)$, suppose $v_i, v_j \in \bar{S}$, where $i \neq j$. Then v_i and v_j are not adjacent, since otherwise, v_i, v_j would form a cycle. Moreover, $\omega(e_{kk}) \neq \lambda$ for all $v_k \in \bar{S}$ by Definition 2.2.

Conversely, suppose that no two vertices $v_i, v_j \in \bar{S}$ are adjacent. Assuming that $v_1, \ldots, v_m$ for $m \geq 2$ is a cycle of G, then if it is the case that $v_1 \in \bar{S}$, it follows that $v_2 \in S$, since v_1 and v_2 are adjacent. Otherwise, if $v_1 \in S$, then in either case, the cycle $v_1, \ldots, v_m$ will contain a vertex of S. Additionally, if $\omega(e_{kk}) \neq \lambda$ for all $v_k \in \bar{S}$, then it follows that S is a structural set of G. □

Based on Theorem 2.3, the set $S = \{v_1, v_3, v_4, v_6\}$ of the graph G in Fig. 2.5 (left) is a structural set, since G has no loops and v_2, v_5 are not adjacent.

2.3 Sequential Graph Reductions

Observe that if $G \in \mathbb{G}$, then every reduction $\mathcal{R}_S(G)$ is again a graph in $\mathbb{G}$. Since $\mathcal{R}_S(G)$ may have a structural set T, it may be possible to consider a sequence of reductions on the graph $G \in \mathbb{G}$. However, to do so formally requires that we first extend our notation to sequences of isospectral graph reductions.

For $G = (V, E, \omega)$, suppose $S_m \subset S_{m-1} \subset \cdots \subset S_1 \subset V$ such that $S_1 \in st(G)$, $\mathcal{R}_1(G) = \mathcal{R}_{S_1}(G)$, and

$$S_{i+1} \in st(\mathcal{R}_i(G)) \text{ where } \mathcal{R}_{S_{i+1}}(\mathcal{R}_i(G)) = \mathcal{R}_{i+1}(G), \ 1 \leq i \leq m-1.$$

If this is the case, we say that $S_1, \ldots, S_m$ *induce a sequence of reductions* on G with *final vertex set* S_m. By way of notation, we write $\mathcal{R}_m(G) = \mathcal{R}(G; S_1, \ldots, S_m)$,

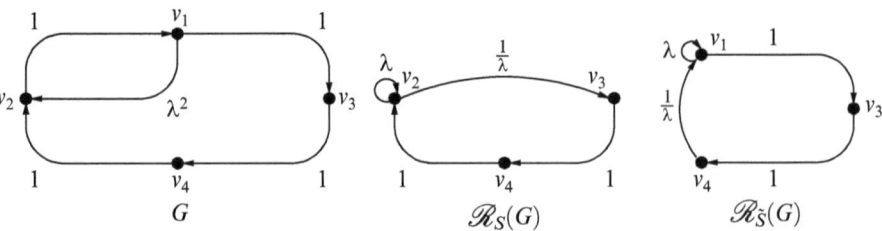

Fig. 2.6 The graph G and its reductions considered in Example 2.5

where $\mathcal{R}(G; S_1, \ldots, S_m)$ denotes the graph G reduced over the vertex set S_1, then over S_2, and so on until G is reduced over the final vertex set S_m.

Because an isospectral graph reduction is a special type of matrix reduction, Theorem 2.1 has the following corollary.

Corollary 2.3. *If $S_1, \ldots, S_m$ induces a sequence of reductions on $G \in \mathbb{G}$, then*

$$M(\mathcal{R}(G; S_1, \ldots, S_m)) = \mathcal{R}(M(G); S_m).$$

If $S \notin st(G)$, it is natural to ask whether there exists a sequence of vertex sets $S \subset S_{m-1} \subset \cdots \subset S_1 \subset V$ that induce a sequence of reductions on G, and if such a sequence exists, whether it is the only such sequence.

To address these questions we consider the following example.

Example 2.5. Consider the graph $G = (V, E, \omega) \in \mathbb{G}$ shown in Fig. 2.6 (left) with adjacency matrix

$$M(G) = \begin{bmatrix} 0 & \lambda^2 & 1 & 0 \\ 1 & 0 & 0 & 0 \\ 0 & 0 & 0 & 1 \\ 0 & 1 & 0 & 0 \end{bmatrix}.$$

Here, our goal is to remove the vertices v_1 and v_2 from G. However, since the set $\{v_3, v_4\}$ is not in $st(G)$ our only option is to remove these vertices sequentially, one at a time.

Removing v_1 then v_2 amounts to finding the isospectral reduction $\mathcal{R}(G; S, T)$, where $S = \{v_2, v_3, v_4\}$ and $T = \{v_3, v_4\}$. Since the set $S = \{v_2, v_3, v_4\}$ is a structural set of G, it is possible to find the reduction $\mathcal{R}(G; S)$, which is shown in Fig. 2.6 (center). However, $\mathcal{R}(G; S)$ cannot be reduced over the vertex set T, since the weight of v_2 in this graph is equal to λ.

Suppose we attempt to reverse the order in which we reduce G by first removing v_2, and then v_1. To do so, we let $\tilde{S} = \{v_1, v_3, v_4\}$, where we attempt to compute the reduction $\mathcal{R}(G; \tilde{S}, T)$. However, the reduction $\mathcal{R}(G; \tilde{S})$ shown in Fig. 2.6 (right) has the same problem as before. Namely, the vertex set T is not a structural set of

2.3 Sequential Graph Reductions

$\mathscr{R}(G; \tilde{S})$. Since we cannot remove v_1 and v_2 together or sequentially from G, we are forced to conclude that it is not possible to reduce G isospectrally to a graph with vertex set T.

To determine when it is possible to reduce a graph $G = (V, E, \omega)$ and when it is not, we note the following. If the weight $\omega(e_{ii})$ is not equal to λ for some $v_i \in V$, then the vertex set $S = V - \{v_i\}$ is a structural set of G. This follows from the fact that $\tilde{S} = \{v_i\}$. Hence, the graph $G|_{\tilde{S}}$ is the graph restricted to the single vertex v_i, and every cycle of $G|_{\tilde{S}}$ is a loop. Therefore, $G \in \mathbb{G}$ can be reduced over the structural set $S = V - \{v_i\}$ if it is known that $\omega(e_{ii}) \neq \lambda$.

Another way to state this is to say that it is possible to remove the vertex v_i from G via an isospectral graph reduction if $\omega(e_{ii}) \neq \lambda$ even when nothing is known about the graph structure of G.

This has the following important implication. Suppose it is known that no loop of G and no loop of any sequential reduction of G has weight λ. If this is the case, then it is possible to remove any sequence of single vertices from G via some sequence of isospectral reductions. Consequently, the graph G can be reduced to a graph on any subset of its vertex set by sequentially removing any "unwanted" vertices.

In general, though, there is no way to know beforehand whether a set of vertices can be removed from a graph without actually going through the process of trying to remove those vertices. However, as with matrices, there is a way to overcome this problem.

Definition 2.6. Let $\mathbb{G}_\pi$ be the set of graphs given by

$$\mathbb{G}_\pi = \{G \in \mathbb{G} : M(G) \in \mathbb{W}_\pi^{n \times n} \text{ for some } n \in \mathbb{N}\}.$$

Lemma 2.1. *If $G \in \mathbb{G}_\pi$ and $S \in st(G)$, then $\mathscr{R}_S(G) \in \mathbb{G}_\pi$. In particular, no loop of G and no loop of any reduction of G can have weight λ.*

Proof. Suppose $G = (V, E.\omega) \in \mathbb{G}_\pi$ has the structural set S. Hence, $M(G) \in \mathbb{W}_\pi^{n \times n}$ for some $n \in \mathbb{N}$. Using Theorems 2.1 and 1.2, it follows that

$$M(\mathscr{R}_S(G)) = \mathscr{R}(M(G); S) \in \mathbb{W}_\pi^{|S| \times |S|},$$

implying $\mathscr{R}_S(G) \in \mathbb{G}_\pi$. Since each edge weight of G and $\mathscr{R}_S(G)$ belongs to $\mathbb{W}_\pi$, it follows that no edge of these graphs can have the weight $\lambda/1 \notin \mathbb{W}_\pi$. This completes the proof. □

A direct consequence of Lemma 2.1 is that if $G \in \mathbb{G}_\pi$, then G can be (sequentially) reduced to a graph on any subset of its vertex set. This result is stated in the following theorem.

Theorem 2.4 (Existence of Sequential Graph Reductions). *Suppose $G = (V, E, \omega)$ is in $\mathbb{G}_\pi$. If $S \subset V$ is nonempty, then there are vertex sets $S \subset S_{m-1} \subset \cdots \subset S_1 \subset V$ such that $S_1, \ldots, S_{m-1}, S$ induces a sequence of reductions on G.*

Proof. Let $G = (V, E, \omega)$ be in $\mathbb{G}_\pi$. If $S \subset V$ is nonempty, then suppose, without loss of generality, that $S = \{v_1, \ldots, v_{n-m}\}$, where $|V| = n$. Letting $S_i = \{v_1, \ldots, v_{n-i}\}$, we see that for each $1 \leq i < n - m$, the claim is that vertex set S_i is in $st(\mathscr{R}_{S_{i+1}}(G))$.

Since S_i and S_{i+1} differ by a single vertex v_{i+1}, it follows that S_i is a structural set of $\mathscr{R}_{S_{i+1}}(G)$, as long as the weight of the loop $e_{i+1,i+1}$ in $\mathscr{R}_{S_{i+1}}(G)$ is not equal to $\lambda \in \mathbb{W}$. Under the assumption that $G \in \mathbb{G}_\pi$, repeated use of Lemma 2.1 implies that this is not the case for each $1 \leq i < n - m$, verifying the claim. Therefore, the sets $S_1, \ldots, S_{m-1}, S$ induce a sequence of reductions on G. □

It is therefore possible to reduce a graph $G \in \mathbb{G}_\pi$ to a graph on any (nonempty) subset of its vertex set via some sequence of isospectral reductions. Observe that Theorem 2.4 does not apply to the graph in Example 2.5, since that graph does not have all weights in the set $\mathbb{W}_\pi$.

Theorem 2.4 can be thought of as an existence result for sequences of isospectral graph reductions. That is, assuming that S is a vertex subset of $G \in \mathbb{G}_\pi$, there is always a sequence of reductions on G with final vertex S. Sequences of reductions also have the following uniqueness property.

Theorem 2.5 (Uniqueness of Sequential Graph Reductions). *Suppose the graph $G = (V, E, \omega)$ is in $\mathbb{G}_\pi$. If $S \subset V$ is nonempty, where both $S_1, \ldots, S_{m-1}, S$ and $T_1, \ldots, T_{n-1}, S$ induce a sequence of reductions on G, then*

$$\mathscr{R}(G; S_1, \ldots, S_{m-1}, S) = \mathscr{R}(G; T_1, \ldots, T_{n-1}, S).$$

Proof. Suppose $G = (V, E, \omega) \in \mathbb{G}_\pi$. If both $S_1, \ldots, S_{m-1}, S$ and $T_1, \ldots, T_{n-1}, S$ induce a sequence of reductions on G, then

$$M(\mathscr{R}(G; S_1, \ldots, S_{m-1}, S)) = \mathscr{R}(M(G); S) = M(\mathscr{R}(G; T_1, \ldots, T_{n-1}, S)),$$

using Corollary 2.3. Since these isospectral graph reductions have the same adjacency matrix, they are the same graph. □

Suppose the graph $G = (V, E, \omega)$ is a graph in $\mathbb{G}_\pi$ and that $S \subset V$ is nonempty. By combining the existence and uniqueness results of Theorems 2.4 and 2.5, it follows that there is exactly one graph that results from sequentially reducing G by any sequence of reductions with final vertex set S. This fact allows us to introduce the following definition.

Definition 2.7. Let $G = (V, E, \omega)$ be a graph in $\mathbb{G}_\pi$. If $S \subseteq V$ is nonempty, define

$$\mathscr{R}_S[G] = \mathscr{R}(G; S_1, \ldots, S_{m-1}, S),$$

where $S_1, \ldots, S_{m-1}, S$ is any sequence that induces a sequence of reductions on G with final vertex set S.

2.3 Sequential Graph Reductions

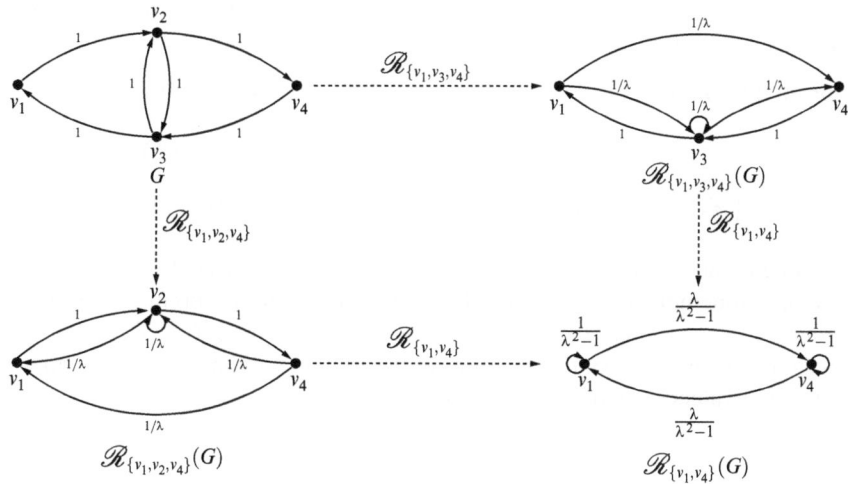

Fig. 2.7 Distinct sequences of isospectral reductions with the same final vertex set and outcome

The graph $\mathcal{R}_S[G]$ is well defined as a result of Theorems 2.4 and 2.5. The notation $\mathcal{R}_S[G]$ given in Definition 2.7 is intended to emphasize the fact that S need not be a structural set of G.

Similar to Remark 1.1, we note that if the adjacency matrix $M(G)$ is in $\mathbb{C}^{n\times n}$, then $G \in \mathbb{G}_\pi$. Therefore, every graph with complex weights can be uniquely reduced over every nonempty subset of its vertex set.

Example 2.6. Let $G = (V, E, \omega)$ be the graph shown in Fig. 2.7. Our goal is to reduce G over the vertex set $\{v_1, v_4\} \subset V$. Note that since $G \in \mathbb{G}_\pi$, Theorem 2.4 guarantees that there is at least one sequence of reductions that reduces G to the graph $\mathcal{R}_{\{v_1,v_4\}}[G]$.

In fact, there are exactly two. This follows from the fact that $\{v_1, v_4\} \notin st(G)$. Hence, G cannot be reduced over $\{v_1, v_4\}$ with a single reduction. However, every (nontrivial) reduction of G removes at least one vertex from G. Therefore, the two possible ways of reducing G over the vertex set $\{v_1, v_4\}$ are

$$\mathcal{R}_{\{v_1,v_4\}}[G] = \mathcal{R}(G; \{v_1, v_2, v_4\}, \{v_1, v_4\}) \text{ and} \tag{2.7}$$

$$\mathcal{R}_{\{v_1,v_4\}}[G] = \mathcal{R}(G; \{v_1, v_3, v_4\}, \{v_1, v_4\}). \tag{2.8}$$

Both of the reductions given in (2.7) and (2.8) are shown in Fig. 2.7. The dashed arrows labeled $\mathcal{R}_T$ in this figure represent the reduction of a graph over some structural set $T \subset V$. This notation is meant to emphasize that this diagram commutes. That is,

$$\mathcal{R}_{\{v_1,v_4\}}\bigl(\mathcal{R}_{\{v_1,v_2,v_4\}}(G)\bigr) = \mathcal{R}_{\{v_1,v_4\}}\bigl(\mathcal{R}_{\{v_1,v_3,v_4\}}(G)\bigr),$$

as guaranteed by Theorem 2.5.

Using Definition 2.7, we state the following general result for isospectral graph reductions, which follows from Theorems 1.1, 2.4, and 2.5.

Theorem 2.6. *If $G = (V, E, \omega) \in \mathbb{G}_\pi$ and $S \subset V$ is nonempty, then*

$$\sigma\big(\mathscr{R}_S[G]\big) = \big(\sigma(G) \cup \sigma^{-1}(G|_{\bar{S}})\big) - \big(\sigma(G|_{\bar{S}}) \cup \sigma^{-1}(G)\big); \text{ and}$$

$$\sigma^{-1}\big(\mathscr{R}_S[G]\big) = \big(\sigma(G|_{\bar{S}}) \cup \sigma^{-1}(G)\big) - \big(\sigma(G) \cup \sigma^{-1}(G|_{\bar{S}})\big).$$

Theorem 2.6 can be thought of as a generalization of Corollary 2.1 from structural sets to any nonempty vertex subset of a graph. However, Theorem 2.6 holds only for graphs in $\mathbb{G}_\pi$, whereas Corollary 2.1 holds for every graph $G \in \mathbb{G}$.

In the following section, the results established here will be used to give rules for devising equivalence relations on the set of graphs $\mathbb{G}_\pi$.

2.4 Equivalence Relations

Theorems 2.4 and 2.5 from the previous section assert that a graph $G \in \mathbb{G}_\pi$ has a unique reduction to any (nonempty) subset of its vertex set via some sequence of isospectral reductions. In this section, this fact will allow us to define equivalence relations on the collection of graphs in $\mathbb{G}_\pi$. To define these equivalence relations, we need first to introduce the notion of isomorphic graphs.

Two weighted digraphs $G_1 = (V_1, E_1, \omega_1)$ and $G_2 = (V_2, E_2, \omega_2)$ are *isomorphic* if there is a bijection $b : V_1 \to V_2$ such that there is an edge e_{ij} in G_1 from v_i to v_j if and only if there is an edge $\tilde{e}_{ij}$ between $b(v_i)$ and $b(v_j)$ in G_2 with $\omega_2(\tilde{e}_{ij}) = \omega_1(e_{ij})$. If the map b exists, it is called an *isomorphism*, and we write $G_1 \simeq G_2$.

An isomorphism is essentially a relabeling of the vertices of a graph. Therefore, if two graphs are isomorphic, then their spectra are identical. The equivalent notion for matrices is that $A(\lambda) \in \mathbb{W}^{n \times n}$ is similar to $B(\lambda) \in \mathbb{W}^{n \times n}$ by some perturbation matrix P, i.e., $A(\lambda) = PB(\lambda)P^{-1}$. This notion of being isomorphic, together with the uniqueness and existence of sequential graph reductions, allows us to define the following equivalence relations on the graphs $\mathbb{G}_\pi$.

Theorem 2.7 (Spectral Equivalence). *Suppose that for each graph $G = (V, E, \omega)$ in $\mathbb{G}_\pi$, τ is a rule that selects a unique nonempty subset $\tau(G) \subset V$. Then τ induces an equivalence relation $\sim$ on the set $\mathbb{G}_\pi$, where $G \sim H$ if $\mathscr{R}_{\tau(G)}[G] \simeq \mathscr{R}_{\tau(H)}[H]$.*

Proof. If $G \in \mathbb{G}_\pi$ the set $\tau(G) \subseteq V$ is unique and nonempty, then the graph $\mathscr{R}_{\tau(G)}[G]$ is uniquely determined by the rule τ. Hence, τ induces a unique reduction on $G \in \mathbb{G}_\pi$.

2.4 Equivalence Relations

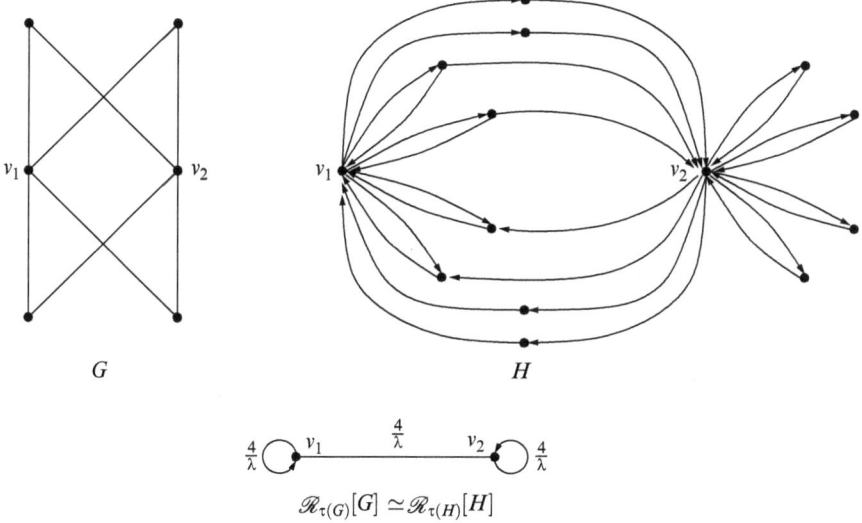

Fig. 2.8 The unweighted undirected graph G is equivalent to the unweighted directed graph H under the relation induced by the rule τ given in Example 2.7. Here, $\tau(G) = \tau(H) = \{v_1, v_2\}$

The claim, then, is that the relation $G \sim H$ if $\mathscr{R}_{\tau(G)}[G] \simeq \mathscr{R}_{\tau(H)}[H]$ is an equivalence relation on $\mathbb{G}_\pi$. This follows from the fact that the relation of being isomorphic is reflexive, symmetric, and transitive, which completes the proof. □

Two graphs may look very different but be spectrally equivalent, suggesting that the corresponding networks have similar dynamics. The major idea in this section is that by choosing an appropriate rule τ, one can discover this similarity.

Example 2.7. Suppose $G = (V, E, \omega)$. For $v \in V$, let $d_{\text{in}}(v)$ be the *in-degree* of v, which is the number of incoming edges incident to v, excluding loops. In an undirected graph, the in-degree of a vertex v is then the same as the number of nonloop edges incident to v.

If $\Delta(G) = \max_{v \in V} d_{\text{in}}(G)$ indicates the *maximum in-degree* of G, let τ be the rule

$$\tau(G) := \{v \in V : d_{\text{in}}(v) > \Delta(G)/2\}.$$

Observe that for each graph $G \in \mathbb{G}_\pi$, the set $\tau(G)$ both exists and is unique. Thus, the relation of having an isomorphic reduction with respect to this rule induces an equivalence relation on $\mathbb{G}_\pi$.

Here, the rule τ selects all those vertices that have an in-degree greater than $\Delta(G)/2$. In Fig. 2.8, the graphs G and H have the vertex set $\tau(G) = \{v_1, v_2\} = \tau(H)$. Moreover, as shown in the figure, the graph $\mathscr{R}_{\tau(G)}[G]$ is isomorphic to $\mathscr{R}_{\tau(H)}[H]$, implying $G \sim H$ under the relation $\sim$ induced by the rule τ. What

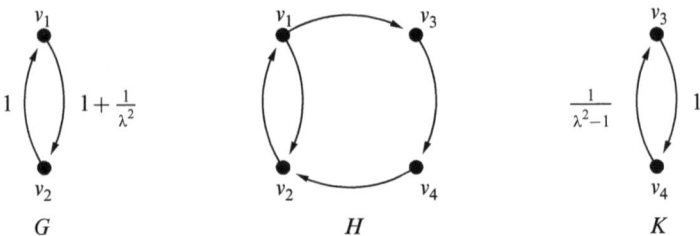

Fig. 2.9 We have $G = \mathscr{R}_S[H]$ and $\mathscr{R}_T[H] = K$ for $S = \{v_1, v_2\}$ and $T = \{v_3, v_4\}$, but the graphs G and K do not have isomorphic reductions

is interesting here is that although these graphs do not appear to have a similar structure, they are the same when reduced with respect to the rule τ.

We note that not every rule that one could propose will select a unique vertex set of a graph. The simplest example is the rule that randomly selects a single vertex of a graph. This selection is, of course, nonunique. The reason it does not lead to an equivalence relation is that without a unique vertex set, the relation cannot be reflexive.

Importantly, choosing a rule that selects a unique vertex set of each graph allows one to study the graphs in $\mathbb{G}_\pi$ modulo some particular graph feature. In Example 2.7, this graph feature, or vertex set, is the set of vertices that have in-degree less than or equal to $\Delta(G)/2$ for every graph G.

From a practical point of view, this allows those studying a particular class of networks a way of comparing the *reduced topology* of these networks. Of course, the particular reduction rule τ should be designed by the particular biologist, chemist, physicist, etc., to have some significance with respect to the networks under consideration.

Before continuing to the next section, we note that the relation of simply having isomorphic reductions is not transitive. That is, if both $\mathscr{R}_S[G] \simeq \mathscr{R}_T[H]$ and $\mathscr{R}_U[H] \simeq \mathscr{R}_V[K]$, it is not necessarily the case that there are sets X and Y, subsets of the vertex sets of G and K respectively, such that $\mathscr{R}_X[G] \simeq \mathscr{R}_Y[K]$.

As an example, in Fig. 2.9, we have both $\mathscr{R}_S[G] \simeq \mathscr{R}_S[H]$ and $\mathscr{R}_T[H] \simeq \mathscr{R}_T[K]$, where $S = \{v_1, v_2\}$ and $T = \{v_3, v_4\}$. However, one can quickly check that for no subsets $X \subseteq S$ and $Y \subseteq T$ do we have $\mathscr{R}_X[G] \simeq \mathscr{R}_Y[K]$. That is, the relation of having isomorphic reductions is not an equivalence relation on $\mathbb{G}$, since it is not transitive. Overcoming this intransitivity requires some rule τ that selects a unique set of vertices from each graph in $\mathbb{G}$ (as in Theorem 2.7).

2.5 Weight-Preserving Isospectral Transformations

Suppose $G = (V, E, \omega)$ is a graph in $\mathbb{G}$. Recall that the *weight set* of G, or the collection of edge weights of G, is the set

$$\omega(E) = \{\omega(e_{ij}) : e \in E\}.$$

The isospectral graph reductions introduced in Sect. 2.2 modify not only the graph structure but also the weight set of a graph. That is, if $\mathcal{R}_S(G) = (S, \mathcal{E}, \mu)$ is a reduction of $G = (V, E, \omega)$, then typically, $\omega(E) \neq \mu(\mathcal{E})$.

This may lead one to assume that the procedure of reducing a graph simply shifts the complexity of the graph's structure to its set of edge weights. While the edge weights can become more complicated in the case of isospectral reductions, this is not the case for the transformations we will consider in this section.

In this section, we introduce graph transformations that modify the structure of a graph but preserve the weights of the graph's edges. As before, this procedure preserves the spectrum of the graph up to a known set of eigenvalues. The idea behind an isospectral graph transformation that preserves a graph's edge weights is based on the following. If two graphs $G, H \in \mathbb{G}$ have the same branch structure (including weights), then they should have similar spectra.

To make this precise, suppose $G = (V, E, \omega)$ and $S \in st(G)$. For the branch $\beta = v_1, \ldots, v_m \in \mathcal{B}_S(G)$, we define $\Omega_G(\beta)$ to be the ordered sequence of weights

$$\Omega_G(\beta) = \omega(e_{12}), \ldots, \omega(e_{i-1,i}), \omega(e_{ii}), \omega(e_{i,i+1}), \ldots, \omega(e_{m-1,m})$$

for $m > 1$, and $\omega(e_{ii})$ if $m = 1$.

Let $G, H \in \mathbb{G}$ and suppose $S = \{v_1, \ldots, v_m\}$ is a structural set of both G and H. We say that the branch set $\mathcal{B}_{ij}(G; S)$ is *isomorphic* to $\mathcal{B}_{ij}(H; S)$ if there is a bijection

$$b : \mathcal{B}_{ij}(G; S) \to \mathcal{B}_{ij}(H; S)$$

such that $\Omega_G(\beta) = \Omega_H(b(\beta))$ for each $\beta \in \mathcal{B}_{ij}(G; S)$.

If such a map exists, we write $\mathcal{B}_{ij}(G; S) \simeq \mathcal{B}_{ij}(H; S)$. If

$$\mathcal{B}_{ij}(G; S) \simeq \mathcal{B}_{ij}(H; S) \text{ for all } 1 \leq i, j \leq m,$$

we say that $\mathcal{B}_S(G)$ is *isomorphic* to $\mathcal{B}_S(H)$ and write $\mathcal{B}_S(G) \simeq \mathcal{B}_S(H)$.

Definition 2.8. For $G, H \in \mathbb{G}$, suppose S is a structural set of both G and H. If

(i) $\mathcal{B}_S(G) \simeq \mathcal{B}_S(H)$, and
(ii) each vertex of G and H belong to a branch of $\mathcal{B}_S(G)$ and $\mathcal{B}_S(H)$ respectively,

we call H is a *weight-preserving isospectral transformation* (wpit) of G over S.

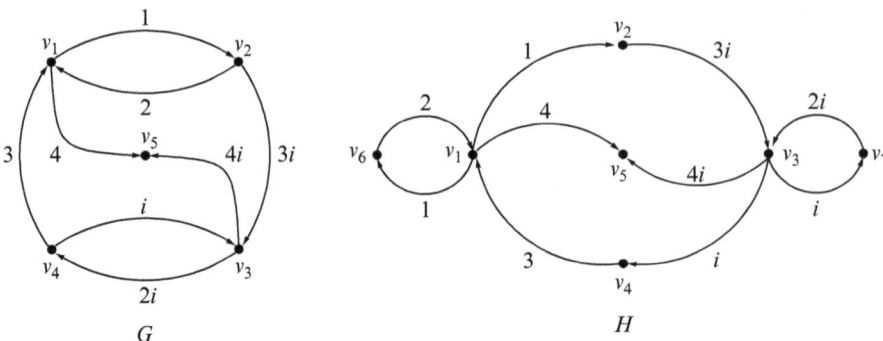

Fig. 2.10 The graph H is a weight-preserving isospectral transformation of the graph G with respect to the structural set $S = \{v_1, v_3, v_5\}$

Notice that if $\mathscr{B}_S(G) \simeq \mathscr{B}_S(H)$ and each vertex of G and H belongs to some branch, then G and H have the same weight set. This justifies the term "weight-preserving" given in Definition 2.8.

More formally, suppose $H = (\mathscr{V}, \mathscr{E}, \mu)$ is a *wpit* of $G = (V, E, \omega)$. By assumption, then, each vertex of G and H belongs to a branch of $\mathscr{B}_S(G)$ and $\mathscr{B}_S(H)$ respectively. Suppose that $e_{ij} \in E$. If $i = j$, then since the vertex v_i belongs to some branch $\beta \in \mathscr{B}_S(G)$, $\omega(e_{ij})$ belongs to the weight sequence $\Omega_G(\beta)$. If $i \neq j$, then there are branches $v_1, \ldots, v_i, \ldots, v_s$ and $u_1, \ldots, u_j, \ldots, u_t$ in $\mathscr{B}_S(G)$ containing v_i and $v_j = u_j$ respectively. Hence, $\beta = v_1, \ldots, v_i, v_j, \ldots, u_t \in \mathscr{B}_S(G)$. Therefore, $\omega(e_{ij})$ belongs to the weight sequence $\Omega_G(\beta)$.

Thus, each edge weight of G belongs to a weight sequence of some $\beta \in \mathscr{B}_S(G)$. By similar reasoning, each edge weight of H belongs to a weight sequence of some $\beta \in \mathscr{B}_S(H)$. Since $\mathscr{B}_S(G) \simeq \mathscr{B}_S(H)$, then $\omega(e_{ij})$ is an edge weight of G if and only if $\omega(e_{ij})$ is an edge weight of H. Therefore, G and H have the same set of edge weights.

Example 2.8. Suppose $G = (V, E, \omega)$ and $H = (\mathscr{V}, \mathscr{E}, \mu)$ are the graphs shown in Fig. 2.10. We note that the vertex set $S = \{v_1, v_3, v_5\}$ is a structural set of both G and H. Moreover,

$$\mathscr{B}_{11}(G; S) = \{v_1, v_2, v_1\}, \quad \mathscr{B}_{11}(H; S) = \{v_1, v_6, v_1\};$$
$$\mathscr{B}_{13}(G; S) = \{v_1, v_2, v_3\}, \quad \mathscr{B}_{13}(H; S) = \{v_1, v_2, v_3\};$$
$$\mathscr{B}_{15}(G; S) = \{v_1, v_5\}, \quad \mathscr{B}_{15}(H; S) = \{v_1, v_5\};$$
$$\mathscr{B}_{31}(G; S) = \{v_3, v_4, v_1\}, \quad \mathscr{B}_{31}(H; S) = \{v_3, v_4, v_1\};$$
$$\mathscr{B}_{33}(G; S) = \{v_3, v_4, v_3\}, \quad \mathscr{B}_{33}(H; S) = \{v_3, v_7, v_3\};$$
$$\mathscr{B}_{35}(G; S) = \{v_3, v_5\}, \quad \mathscr{B}_{35}(H; S) = \{v_3, v_5\};$$
$$\mathscr{B}_{51}(G; S) = \emptyset, \quad \mathscr{B}_{51}(H; S) = \emptyset;$$

2.5 Weight-Preserving Isospectral Transformations

$$\mathcal{B}_{53}(G;S) = \emptyset, \quad \mathcal{B}_{53}(H;S) = \emptyset;$$
$$\mathcal{B}_{55}(G;S) = \emptyset, \quad \mathcal{B}_{55}(H;S) = \emptyset.$$

Hence, there is a bijection $b : \mathcal{B}_{ij}(G;S) \to \mathcal{B}_{ij}(H;S)$ for all $i, j \in \{1, 3, 5\}$.

To determine whether the branch sets $\mathcal{B}_S(G)$ and $\mathcal{B}_S(H)$ are isomorphic, we need to check whether each branch $\beta \in \mathcal{B}_S(G)$ and the corresponding branch $b(\beta) \in \mathcal{B}_S(H)$ have the same sequence of weights. Observe that the branch $\beta = v_1, v_2, v_1$ in $\mathcal{B}_{11}(G;S)$ has the weight sequence $\Omega_G(\beta) = 1, 0, 2$. Similarly, note that the branch $b(\beta) = v_1, v_6, v_1$ in $\mathcal{B}_{11}(H;S)$ has the weight sequence $\Omega_H(b(\beta)) = 1, 0, 2$. Hence, $\mathcal{B}_{11}(G;S) \simeq \mathcal{B}_{11}(H;S)$. Continuing in this manner, one can check that each $\mathcal{B}_{ij}(G;S)$ is isomorphic to $\mathcal{B}_{ij}(H;S)$ for $i, j \in \{1, 3, 5\}$, so that $\mathcal{B}_S(G) \simeq \mathcal{B}_S(H)$. Since every vertex of G and H belongs to a branch of $\mathcal{B}_S(G)$ or $\mathcal{B}_S(H)$ respectively, H is a *wpit* of G over S.

Observe that the set S is not the only structural set that is common to both G and H. Another structural set of both graphs is the set $T = \{v_1, v_3\}$. For this set, one can again show that $\mathcal{B}_T(G) \simeq \mathcal{B}_T(H)$. However, since the vertex v_5 does not belong to any branch in either $\mathcal{B}_T(G)$ or $\mathcal{B}_T(H)$, it follows that the graph H is not a *wpit* of G over T.

The reason we do not consider H to be a *wpit* of G with respect to the set T is that some edge weights of these graphs are unaccounted for by the branch sets $\mathcal{B}_T(G)$ and $\mathcal{B}_T(H)$. For instance, if $\omega(e_{15}) = \pi/2$, rather than $\omega(e_{15}) = 4$, as shown in Fig. 2.10, then $\omega(E) \neq \mu(\mathcal{E})$, although the branch sets $\mathcal{B}_T(G)$ and $\mathcal{B}_T(H)$ would still be isomorphic. That is, part (ii) of Definition 2.8 is a necessary condition for ensuring that a *wpit* of a graph has the same weights as the original untransformed graph.

Suppose that two graphs $G, H \in \mathbb{G}$ both have the structural set S. If it is the case that $\mathcal{B}_S(G) \simeq \mathcal{B}_S(H)$, then $\mathcal{R}_S(G) = \mathcal{R}_S(H)$. This leads to the following result, which is a corollary of Theorem 2.1.

Corollary 2.4. *If the graph $G \in \mathbb{G}$ is a weight-preserving isospectral transformation of H over S, then*

$$\sigma(\mathcal{R}_S(G)) = G_* - G^* = H_* - H^* \text{ and } \sigma^{-1}(\mathcal{R}_S(G)) = G^* - G_* = H^* - H_*,$$

where $G_* = \sigma(G) \cup \sigma^{-1}(G|_{\bar{S}})$, $G^* = \sigma(G|_{\bar{S}}) \cup \sigma^{-1}(G)$, $H_* = \sigma(H) \cup \sigma^{-1}(H|_{\bar{S}})$, and $H^* = \sigma(H|_{\bar{S}}) \cup \sigma^{-1}(H)$.

Example 2.9. As a demonstration of Corollary 2.4, consider the graphs G and H in Fig. 2.10, which are *wpits* of each other over the set $S = \{v_1, v_3, v_5\}$. Since $M(G)$ and $M(H)$ have complex-valued entries, each of $\sigma^{-1}(G)$, $\sigma^{-1}(G|_{\bar{S}})$, $\sigma^{-1}(H)$, and $\sigma^{-1}(H|_{\bar{S}})$ is the empty set. Hence, Corollary 2.4 implies that

$$\sigma(\mathcal{R}_S(G)) = \sigma(G) - \sigma(G|_{\bar{S}}) = \sigma(H) - \sigma(H|_{\bar{S}}).$$

Indeed, as one can compute, $\sigma(G) = \{\pm(-5)^{\frac{1}{4}}, \pm(-5)^{\frac{1}{4}}i, 0\}$, $\sigma(G|_{\bar{S}}) = \{0, 0\}$, $\sigma(H) = \{\pm(-5)^{\frac{1}{4}}, \pm(-5)^{\frac{1}{4}}i, 0, 0, 0\}$, and $\sigma(H|_{\bar{S}}) = \{0, 0, 0, 0\}$, verifying this fact. Additionally, this implies that

$$\sigma^{-1}(\mathcal{R}_S(G)) = \sigma(G|_{\bar{S}}) - \sigma(G) = \sigma(H|_{\bar{S}}) - \sigma(H),$$

demonstrating the second half of Corollary 2.4.

2.5.1 Branch Expansions

Typically, a graph $G \in \mathbb{G}$ will have many weight-preserving isospectral transformations over some $S \in st(G)$. Because G and each of these transformations have the same branch structure, there is some largest isospectral transformation of G. This is the transformation H in which the branches $\mathcal{B}_S(H)$ overlap the least. We call such a *wpit* the *isospectral expansion* of G.

More precisely, suppose $\alpha = v_1, \ldots, v_m$ and $\beta = u_1, \ldots, u_n$ are branches in $\mathcal{B}_S(G)$. Then these branches are said to be *independent* if

$$\{v_2, \ldots, v_{m-1}\} \cap \{u_2, \ldots, u_{n-1}\} = \emptyset.$$

That is, α and β are independent if they share no interior vertices.

Definition 2.9. Let $G, H \in \mathbb{G}$ and $S \in st(G), st(H)$. Suppose

(i) H is a weight-preserving isospectral transformation of G over S, and
(ii) the branches of $\mathcal{B}_S(H)$ are independent.

Then we call H an *isospectral expansion* of G with respect to S.

Isospectral expansions are particular types of weight-preserving isospectral transformations and moreover, are unique up to a labeling of vertices. That is, every two isospectral expansions of G with respect to S are isomorphic. By slight abuse of terminology, we let $\mathcal{X}_S(G)$ be any representative from the set of isospectral expansions and call $\mathcal{X}_S(G)$ the *isospectral expansion* of G with respect to S.

Since an isospectral expansion transforms a graph in a very specific way, it is possible to relate the spectrum of a graph with its expansion. This relation is given in the following theorem.

Theorem 2.8. *Let $G = (V, E, \omega)$ with structural set S. Then*

$$\det(M(\mathcal{X}_S(G)) - \lambda I) = \det(M(G) - \lambda I) \prod_{v_i \in V-S} (\omega(e_{ii}) - \lambda)^{n_i - 1},$$

where n_i is the number of branches in $\mathcal{B}_S(G)$ containing v_i.

2.5 Weight-Preserving Isospectral Transformations

Proof. Let $\mathscr{X}_S(G) = (\mathscr{V}, \mathscr{E}, \mu)$ be an isospectral expansion of the graph $G = (V, E, \omega)$. Suppose $\beta = v_1, \ldots, v_\ell \in \mathscr{B}_S(\mathscr{X}_S(G))$. Since $\mathscr{B}_S(G) \simeq \mathscr{B}_S(\mathscr{X}_S(G))$, there is a bijection

$$b : \mathscr{B}_S(G) \to \mathscr{B}_S(\mathscr{X}_S(G))$$

such that $\Omega_{\mathscr{X}_S(G)}(\beta) = \Omega_G(b^{-1}(\beta))$.

If $\beta(i) = v \notin S$, define

$$\tilde{b}(v_i) = (b^{-1}(\beta))(i) \text{ for } 1 < i < \ell.$$

Since $|\beta| = |b^{-1}(\beta)|$, then $\tilde{b}$ maps interior vertices of β to interior vertices of $b^{-1}(\beta)$. Since each vertex of $\mathscr{V} - S$ and $V - S$ belong to some branch of $\mathscr{B}_S(\mathscr{X}_S(G))$ and $\mathscr{B}_S(G)$ respectively, it follows that $\tilde{b} : \mathscr{V} - S \to V - S$ is onto.

Let

$$\mathscr{V}_i = \{v \in \mathscr{V} - S : \tilde{b}(v) = v_i \in V\}.$$

Note that $|\mathscr{V}_i| = n_i$ is then the number of branches in $\mathscr{B}_S(G)$ containing v_i. Moreover, $\mathscr{V} - S$ is the disjoint union

$$\mathscr{V} - S = \bigcup_{v_i \in V - S} \mathscr{V}_i.$$

Since $b : \mathscr{B}_S(G) \to \mathscr{B}_S(\mathscr{X}_S(G))$ preserves weight sequences, we have $\mu(e_{jj}) = \omega(e_{ii})$ for each $v_j \in \mathscr{V}_i$. Hence,

$$\prod_{v_j \in \mathscr{V} - S} (\mu(e_{jj}) - \lambda) = \prod_{v_i \in V - S} \left(\prod_{v_j \in \mathscr{V}_i} (\mu(e_{jj}) - \lambda) \right) = \prod_{v_i \in V - S} (\omega(e_{ii}) - \lambda)^{n_i}. \tag{2.9}$$

If $S = \{v_1, \ldots, v_m\}$, then by assumption,

$$\mathscr{B}_{ij}(G; S) \simeq \mathscr{B}_{ij}(\mathscr{X}_S(G); S) \text{ for } 1 \leq i, j \leq m.$$

Equation (2.2) then implies that the edges e_{ij} in both $\mathscr{R}_S(G)$ and $\mathscr{R}_S(\mathscr{X}_S(G))$ have the same weight. Therefore, $\mathscr{R}_S(G) = \mathscr{R}_S(\mathscr{X}_S(G))$. Since

$$\det(M(\mathscr{R}_S(G)) - \lambda I) = \det(M(\mathscr{R}_S(\mathscr{X}_S(G))) - \lambda I),$$

equation (1.11), used in the context of the adjacency matrices of G and $\mathscr{X}_S(G)$, implies

$$\frac{\det\left(M(G) - \lambda I\right)}{\det\left(M(G|_{\bar{S}}) - \lambda I\right)} = \frac{\det\left(M(\mathscr{X}_S(G)) - \lambda I\right)}{\det\left(M(\mathscr{X}_S(G)|_{\bar{S}}) - \lambda I\right)}.$$

Equation (2.5) then implies that

$$\det\left(M(\mathscr{X}_S(G)) - \lambda I\right) = \det\left(M(G) - \lambda I\right) \frac{\prod_{v_j \in \mathscr{V}-S}\left(\mu(e_{jj}) - \lambda\right)}{\prod_{v_i \in V-S}\left(\omega(e_{ii}) - \lambda\right)}.$$

From (2.9) we then have that

$$\det\left(M(\mathscr{X}_S(G)) - \lambda I\right) = \det\left(M(G) - \lambda I\right) \prod_{v_i \in V-S}\left(\omega(e_{ii}) - \lambda\right)^{n_i - 1},$$

completing the proof. □

Example 2.10. Consider the graphs $G = (V, E, \omega)$ and $H = (\mathscr{V}, \mathscr{E}, \mu)$ in Fig. 2.10. As demonstrated in Example 2.8, if $S = \{v_1, v_3, v_5\}$, then the graph H is a *wpit* of G. Moreover, it can be seen from Fig. 2.10 that the six branches of $\mathscr{B}_S(H)$ share no interior vertices. That is, the branches of $\mathscr{B}_S(H)$ are pairwise independent.

Therefore, the graph H is an isospectral expansion of G with respect to S or $H = \mathscr{X}_S(G)$. Importantly, note that the edge weights of G and its expansion H are identical, i.e., as sets, $\omega(E) = \mu(\mathscr{E}) = \{1, 2, 3, 4, i, 2i, 3i, 4i\}$.

Moreover, since $V - S = \{v_2, v_4\}$, observe that the vertex v_2 of G is an interior vertex of the branches v_1, v_2, v_1; $v_1, w_2, v_3 \in \mathscr{B}_S(G)$. Similarly, the vertex v_4 of G is an interior vertex of the branches v_3, v_4, v_3; $v_3, v_4, v_1 \in \mathscr{B}_S(G)$. Since $\omega(e_{22}) = 0$ and $\omega(e_{44}) = 0$, Theorem 2.8 implies that $\sigma(\mathscr{X}_S(G)) = \sigma(G) \cup \{0, 0\}$.

The principal idea behind an isospectral expansion is the following. If $G \in \mathbb{G}$ and $S \in st(G)$, then the set of branches $\mathscr{B}_S(G)$ is uniquely defined. However, there are typically many other graphs H with the same branch structure as G, i.e., $S \in st(H)$, such that $\mathscr{B}_S(H) \simeq \mathscr{B}_S(G)$.

An isospectral expansion of G over S is then a graph $H = \mathscr{X}_S(G)$ with identical branch structure but with the following restriction. The branches of $\mathscr{B}_S(H)$ are pairwise independent, and every vertex of H belongs to a branch in $\mathscr{B}_S(H)$. That is, every vertex of $\bar{S}$ in H is part of exactly one branch in $\mathscr{B}_S(H)$.

Hence, given a graph G and structural set S, we can algorithmically construct the expansion $\mathscr{X}_S(G)$ as follows. Start with the vertices S. If $\beta \in \mathscr{B}_{ij}(G; S)$, then both v_i and v_j are in S. Construct a path (or cycle if $i = j$) from v_i to v_j with the weight sequence $\Omega(\beta)$ using "new" interior vertices. By *new*, we mean vertices that do not already appear on the graph we are constructing. Repeat this for each $\beta \in \mathscr{B}_{ij}(G; S)$. The resulting graph is the isospectral expansion $\mathscr{X}_S(G)$.

Importantly, an isospectral expansion is not the only example of an isospectral graph transformation that preserves the weight set of a graph. Many other weight-preserving isospectral transformations are possible.

2.6 Isospectral Graph Transformations over Modified Weight Sets

The isospectral expansions in Sect. 2.5 are a type of graph transformation that separates the various branches of a graph. In this section, we consider the reversal of this process. Specifically, we introduce a method of transforming a graph in a way that merges branches. This type of isospectral transformation has the additional property that it keeps the graph's weight set in a fixed subset $\mathbb{U} \subset \mathbb{W}$. The particular subsets for which this will hold are semirings of $\mathbb{W}$.

The set $\mathbb{U}$ is a *semiring* of $\mathbb{W}$ if it has the following properties: Both 0 and 1 belong to $\mathbb{U}$, and for every $u_1, u_2 \in \mathbb{U}$, both the product $u_1 u_2$ and the sum $u_1 + u_2$ are in $\mathbb{U}$. However, the additive and multiplicative inverses of u_1 and u_2 need not be in $\mathbb{U}$. Examples of such semirings include $\mathbb{C}[\lambda]$, $\mathbb{R}$, and $\mathbb{Z}^+ = \{0, 1, 2, \dots\}$.

In order to merge the branches of a graph G, we will first need to choose a structural set $S \in st(G)$, so that these branches are defined. The particular structural sets that we will consider in this section are defined as follows.

Definition 2.10. For $G = (V, E, \omega)$, let $S \in st(G)$. The set S is a *complete structural set* of G if each cycle of G, including loops, contains a vertex in S.

The difference between a structural set and a complete structural set of a graph G is the following. If S is a complete structural set of G, then every cycle of G contains a vertex in S. If S is simply a structural set of G, then loops of G need not contain a vertex of S.

For $G \in \mathbb{G}$, let $st_0(G)$ denote the set of all complete structural sets of G. Additionally, let $\sigma_0(G)$ be the nonzero elements of $\sigma(G)$ including multiplicities, which we refer to as the *nonzero spectrum* of G. Moreover, we denote by $\rho(G)$ the *spectral radius* of G. That is,

$$\rho(G) = \max_{\ell \in \sigma(G)} |\ell|.$$

Analogously, if $M \in \mathbb{W}^{n \times n}$, we say that $S \subset N$ is a complete structural set of M if there is a permutation matrix $P \in \mathbb{C}^{|\bar{S}| \times |\bar{S}|}$ such that the matrix $PM_{\bar{S}\bar{S}}P^{-1}$ is upper triangular and each diagonal entry is zero. Let $st_0(M)$ be the set of all complete structural sets of M, $\sigma_0(M)$ the nonzero eigenvalues of M, and $\rho(M)$ the spectral radius of M.

Note that if S is a complete structural set of a graph G, then $G|_{\bar{S}}$ has no cycles. Hence, one can show that

$$\det(M(G|_{\bar{S}}) - \lambda I) = \lambda^{|\bar{S}|}.$$

The following is then a corollary of Theorems 2.1 and 2.8.

Corollary 2.5. Let $G = (V, E, \omega)$ and suppose $S \in st_0(G)$. Then $\sigma_0(\mathscr{R}_S(G)) = \sigma_0(G) = \sigma_0(\mathscr{X}_S(G))$ and $\rho(\mathscr{R}_S(G)) = \rho(G) = \rho(\mathscr{X}_S(G))$.

Proof. Suppose $G = (V, E, \omega)$ and $S \in st_0(G)$. Since no vertex in $\bar{S}$ can have a loop, it follows that $\omega(e_{ii}) = 0$ for each $v_i \in \bar{S}$. Equation (2.5) then implies that

$$\det(M(G|_{\bar{S}}) - \lambda I) = \prod_{v_i \in \bar{S}} \lambda.$$

Hence $\sigma^{-1}(G|_{\bar{S}}) = \emptyset$ and $\sigma(G|_{\bar{S}}) = \{0, \ldots, 0\}$, in which 0 has multiplicity $|\bar{S}|$. The corollary then follows from Theorem 2.1. □

To phrase Corollary 2.5 in terms of matrices, suppose $M \in \mathbb{W}^{n \times n}$ and $S \in st_0(M)$. Then $\sigma_0(\mathscr{R}_S(M)) = \sigma_0(M) = \sigma_0(\mathscr{X}_S(M))$ and $\rho(\mathscr{R}_S(M)) = \rho(M) = \rho(\mathscr{X}_S(M))$.

In what follows, we demonstrate how the branches $\mathscr{B}_S(G)$ of a graph can be merged if $S \in st_0(G)$. To simplify the discussion, we will say the $e_{j,j+1}$ is the jth edge *belonging* to the branch β for each $1 \leq j \leq m - 1$. This allows us to state the following result.

Lemma 2.2. *Let $G = (V, E, \omega)$ and $S \in st_0(G)$. Then $\mathscr{X}_S(G) = (\mathscr{V}, \mathscr{E}, \mu)$ has the following properties:*

(i) *If $e_{ij} \in \mathscr{E}$, then e_{ij} belongs to exactly one branch of $\mathscr{B}_S(\mathscr{X}_S(G))$.*
(ii) *If $\beta = v_1, \ldots, v_m$ is a branch in $\mathscr{B}_S(\mathscr{X}_S(G))$, then*

$$\Omega_{\mathscr{X}_S(G)}(\beta) = \mu(e_{12}), 0, \ldots, \mu(e_{k,k-1}), 0, \mu(e_{k,k+1}), \ldots, 0, \mu(e_{m-1,m}). \tag{2.10}$$

Proof. Since each vertex of $\mathscr{X}_S(G)$ belongs to at least one branch of $\mathscr{B}_S(\mathscr{X}_S(G))$, the same holds for every $e_{ij} \in \mathscr{E}$. On the other hand, suppose e_{ij} belongs to both β_1 and β_2 in $\mathscr{B}_S(\mathscr{X}_S(G))$. Then neither v_i nor v_j can be an interior vertex of β_1 or β_2, since these branches, if distinct, are independent. Hence $\beta_1 = \beta_2 = v_i, v_j$ or e_{ij} belongs to at most one branch of $\mathscr{B}_S(\mathscr{X}_S(G))$. This verifies property (i).

Since S is a complete structural set of G, it follows that $\omega(e_{ii}) = 0$ for each $v_i \in V - S$. This implies that for every $\beta = v_1, \ldots, v_m \in \mathscr{B}_S(G)$, the weight sequence $\Omega_G(\beta)$ has the form given in equation (2.10). Since $\mathscr{B}_S(G) \simeq \mathscr{B}_S(\mathscr{X}_S(G))$, property (ii) holds. □

2.6.1 Branch Reweighting

Given an isospectral expansion $\mathscr{X}_S(G)$, our goal is to construct a new graph $\mathscr{Y}_S(G)$ by reweighting the branches of $\mathscr{X}_S(G)$. The idea behind this construction is to reweight the branches $\mathscr{B}_S(\mathscr{X}_S(G))$ in such a way that it preserves their branch products.

Suppose the expansion $\mathscr{X}_S(G)$ is equal to $(\mathscr{V}, \mathscr{E}, \mu)$, where $S \in st_0(G)$. Let the graph $\mathscr{Y}_S(G)$ be equal to $(\mathscr{V}, \mathscr{E}, \nu)$. That is, $\mathscr{Y}_S(G)$ has the same vertex and edge

2.6 Isospectral Graph Transformations over Modified Weight Sets

set as $\mathcal{X}_S(G)$ but possibly different edge weights. This implies that S is a complete structural set of $\mathcal{Y}_S(G)$ and moreover, that the branch set $\mathcal{B}_S(\mathcal{Y}_S(G))$ is identical to $\mathcal{B}_S(\mathcal{X}_S(G))$.

For the branch $\beta = v_1, \ldots, v_m \in \mathcal{B}_S(\mathcal{Y}_S(G))$, let β have the weight sequence

$$\Omega_{\mathcal{Y}_S(G)}(\beta) = \prod_{k=1}^{m-1} \mu(e_{k,k+1}), 0, \ldots, 1, 0, 1, \ldots, 0, 1 \qquad (2.11)$$

if $m > 1$. If $m = 1$, let $\Omega_{\mathcal{Y}_S(G)}(\beta) = \mu(e_{11})$. Since $\mathcal{X}_S(G)$ and $\mathcal{Y}_S(G)$ have the same vertex and edge sets, Lemma 2.2 implies that each edge of $\mathcal{Y}_S(G)$ belongs to exactly one branch of $\mathcal{B}_S(\mathcal{Y}_S(G))$. Therefore, equation (2.11) completely specifies the edge weights of the graph $\mathcal{Y}_S(G)$.

Observe that the edge $e_{ij} \in \mathcal{E}$ in $\mathcal{Y}_S(G)$ has weight $\nu(e_{ij}) = 1$ unless e_{ij} is the first edge of a branch in $\mathcal{B}_S(\mathcal{Y}_S(G))$. If e_{ij} happens to be the first edge of the branch $\beta \in \mathcal{B}_S(\mathcal{Y}_S(G))$, then its weight is the product of the nonzero entries of $\Omega_{\mathcal{X}_S(G)}(\beta)$.

In effect, the branch reweighting process simply transfers the entire weight of the branch to the branch's first edge, leaving all other edges with unit weight. Hence, the branch product remains constant, but the entire weight of the branch is concentrated on its first edge. (Figure 2.11 gives an example of this branch reweighting.)

2.6.2 Branch Merging

From $\mathcal{Y}_S(G)$, we construct the graph $\mathcal{Z}_S(G)$. The major idea behind this construction is that the branches $\mathcal{B}_S(\mathcal{Y}_S(G))$ of the graph $\mathcal{Y}_S(G)$ can be merged together in a way that maintains the weight set of the graph as well as its nonzero spectrum. To define the graph $\mathcal{Z}_S(G)$, we will need the following terminology.

For $G = (V, E, \omega)$, suppose $S = \{v_1, \ldots, v_m\}$ is a structural set of G. Let

$$\mathcal{B}_j(G; S) = \bigcup_{1 \leq i \leq m} \mathcal{B}_{ij}(G; S).$$

That is, $\mathcal{B}_j(G; S)$ are the branches in $\mathcal{B}_S(G)$ terminating at the same vertex $v_j \in V$. For every $\beta = u_1, \ldots, u_k$ in $\mathcal{B}_S(G)$, let $|\beta| = k$, i.e., the number of vertices in β. Moreover, let $\{\beta\}_{\text{int}}$ denote the set of interior vertices of β.

To construct the graph $\mathcal{Z}_S(G)$, we first suppose that the graph $\mathcal{Y}_S(G)$ is equal to $(\mathcal{V}, \mathcal{E}, \nu)$ and that $S = \{v_1, \ldots, v_m\} \in st_0(G)$. For each $v_j \in S$, we select a branch $\beta^j \in \mathcal{B}_j(\mathcal{Y}_S(G))$ with the property that $|\beta^j| \geq |\beta|$ for all branches $\beta \in \mathcal{B}_S(\mathcal{Y}_S(G))$. That is, the branch β^j is the "longest" branch that terminates at v_j. If $\mathcal{B}_j(\mathcal{Y}_S(G)) = \emptyset$, we set $\beta^j = \emptyset$.

Define $B = \mathscr{B}_S(\mathscr{Y}_S(G)) - \{\beta^1, \ldots, \beta^m\}$, and let

$$\mathscr{U} = \bigcup_{\beta \in B} \{\beta\}_{\text{int}}. \tag{2.12}$$

That is, $\mathscr{U}$ is the set of interior vertices of the branches $\beta \neq \{\beta^1, \ldots, \beta^m\}$ in $\mathscr{B}_S(\mathscr{Y}_S(G))$.

Set

$$\mathscr{L}'_S(G) = \mathscr{Y}_S(G)|_{\mathscr{V} - \mathscr{U}}.$$

Recall that the branches of $\mathscr{B}_S(\mathscr{Y}_S(G))$ are mutually independent. Therefore, the graph $\mathscr{L}'_S(G)$ is equal to $(\mathscr{V} - \mathscr{U}, \mathscr{E}', v')$, where $e \in \mathscr{E}'$ if and only if e belongs to some β^j. Furthermore, the edge weights of $\mathscr{L}'_S(G)$ are given by the restriction $v' = v|_{\mathscr{E}'}$.

If e is an edge from the vertex a to the vertex b, we will denote this by $e = (a, b)$. Suppose the branch β^j is equal to $v_1^j, \ldots, v_k^j$. For each

$$\beta \in \mathscr{B}_{ij}(\mathscr{Y}_S(G); S) - \beta^j,$$

we add an edge $(v_i, v_{k-|\beta|+2}^j)$ to the graph $\mathscr{L}'_S(G)$. The edge $(v_i, v_{k-|\beta|+2}^j)$ is given the weight of the first edge belonging to β in $\mathscr{Y}_S(G)$. If this is done over all $1 \leq i, j \leq m$, we call the resulting graph $\mathscr{L}''_S(G)$.

It is important to note that the graph $\mathscr{L}''_S(G)$ may have parallel edges under this construction. By *parallel edges*, we mean that there may be multiple edges in the edges set of $\mathscr{L}''_S(G)$ of the form (a, b). In particular, if there are two branches $\beta_1, \beta_2 \in \mathscr{B}_{ij}(\mathscr{Y}_S(G); S)$ that have the same length, i.e., $|\beta_1| = |\beta_2| = \ell$, then there are (at least) two edges in $\mathscr{L}''_S(G)$ of the form $(v_i, v_{\ell-|\beta|+2}^j)$.

If the graph $\mathscr{L}''_S(G)$ has parallel edges $e_1, \ldots, e_N$ of the form (v_i, v_j) with weights $w_1, \ldots, w_N$, we replace the edges $e_1, \ldots, e_N$ in $\mathscr{L}''_S(G)$ with the single edge e_{ij} having weight $w_1 + \cdots + w_N$. If this is done for each set of parallel edges in $\mathscr{L}''_S(G)$, we denote the resulting graph by $\mathscr{L}_S(G)$.

Note that our construction of $\mathscr{L}_S(G)$ depends on the initial choice of each β^j. We therefore write $\mathscr{L}_S(G) = \mathscr{L}_S(G; \beta^1, \ldots, \beta^m)$.

Definition 2.11. If $S \in st_0(G)$, then we call the graph $\mathscr{L}_S(G)$ the *merged graph* of G over S.

Because each operation of expanding, reweighting, and merging does not modify the underlying branch structure of a graph or any of its branch products, the following holds.

2.6 Isospectral Graph Transformations over Modified Weight Sets

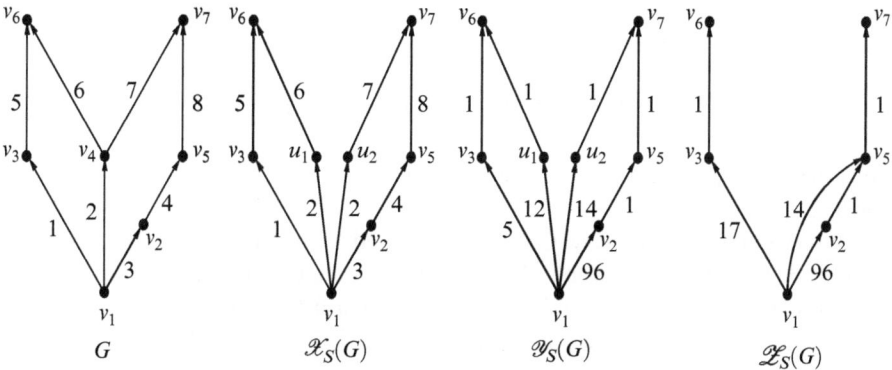

Fig. 2.11 The isospectral transformation $\mathcal{X}_S(G)$ of G over $S = \{v_1, v_2, v_3\}$

Theorem 2.9. *For $G \in \mathbb{G}$ and $S \in st_0(G)$, suppose the edge weights of G are in the semiring $\mathbb{U} \subseteq \mathbb{W}$. Then*

(i) *$\mathcal{X}_S(G)$ has edge weights in the set $\mathbb{U}$;*
(ii) *the nonzero eigenvalues $\sigma_0(\mathcal{X}_S(G))$ are equal to $\sigma_0(G)$; and*
(iii) *the spectral radius $\rho(\mathcal{X}_S(G))$ is equal to $\rho(G)$.*

Proof. The fact that $\mathcal{X}_S(G)$ and G have weights in the semiring $\mathbb{U}$ follows from the fact that the weights of $\mathcal{X}_S(G)$ are sums and products of the weights of G. The nonzero spectrum and spectral radius of $\mathcal{X}_S(G)$ and G follow from Corollary 2.4, since by construction, $\mathcal{R}_S(G) = \mathcal{R}_S(\mathcal{X}_S(G))$. □

Example 2.11. Let $G = (V, E, \omega)$ be the graph shown in Fig. 2.11 (far left). Note that the vertex set $S = \{v_1, v_2, v_3\}$ is a complete structural set of G, since G has no cycles (including loops). The branches of G with respect to S are then $\beta_1 = v_1, v_3, v_6$, $\beta_2 = v_1, v_4, v_6$, $\beta_3 = v_1, v_4, v_7$, and $\beta_4 = v_1, v_2, v_5, v_7$. Observe that only β_2 and β_3 share an interior vertex.

The isospectral expansion $\mathcal{X}_S(G)$, shown in Fig. 2.11 (middle left), is then the graph in which the branches β_2 and β_3 have been replaced with the independent branches $\tilde{\beta}_2 = v_1, u_1, v_6$ and $\tilde{\beta}_3 = v_1, u_2, v_7$ respectively. Thus, the expansion $\mathcal{X}_S(G)$ has the vertex set $\mathcal{V} = \{v_1, v_2, v_3, v_5, v_6, v_7, u_1, u_2\}$.

Note that the products of the weights along the branches $\beta_1, \tilde{\beta}_2, \tilde{\beta}_3, \beta_4$ are 5, 12, 14, 96 respectively. By construction, these are the first edge weights of each of the branches $\beta_1, \tilde{\beta}_2, \tilde{\beta}_3, \beta_4$ in $\mathcal{Y}_S(G)$, respectively (see Fig. 2.11, middle right). Each other edge of $\mathcal{Y}_S(G)$ has unit weight.

To construct $\mathcal{X}'_S(G)$, note that $\beta^1 = \emptyset$, $\beta^7 = \beta_4$, and β^6 is either β_1 or $\tilde{\beta}_2$. Here we make the arbitrary choice of letting $\beta^6 = \beta_1$. Since $\{\tilde{\beta}_2\}_{\text{int}} = \{u_1\}$ and $\{\tilde{\beta}_3\}_{\text{int}} =$

$\{u_2\}$, we have that $\mathscr{L}'_S(G)$ has vertex set $\mathscr{V} - \{u_1, u_2\} = \{v_1, v_2, v_3, v_5, v_6, v_7\}$. Given that the branch set

$$\mathscr{B}_S(\mathscr{L}'_S(G)) - \{\beta^6, \beta^7\} = \{\tilde{\beta}_2, \tilde{\beta}_3\},$$

constructing $\mathscr{L}''_S(G)$ amounts to adding two edges to $\mathscr{L}'_S(G)$, one for $\tilde{\beta}_2$ and one for $\tilde{\beta}_3$.

Observe that $\tilde{\beta}_2 \in \mathscr{B}_{16}(\mathscr{Y}_S(G); S)$ with first edge weight 12 and $|\tilde{\beta}_2| = 3$. Since $|\beta^6| = 3$, we add to the graph $\mathscr{L}'_S(G)$ a parallel edge from v_1 to v_3 with weight 12. For the branch $\tilde{\beta}_3 \in \mathscr{B}_{17}(\mathscr{Y}_S(G); S)$, note that it has first edge weight 14 and $|\tilde{\beta}_2| = 3$. Since $|\beta^7| = 4$, we add to the graph $\mathscr{L}'_S(G)$ an edge from v_1 to v_5 with weight 14. The result is the graph $\mathscr{L}''_S(G)$.

Note that the two parallel edges from v_1 to v_3 in $\mathscr{L}''_S(G)$ have weights 5 and 12 respectively. Hence, in $\mathscr{L}_S(G) = \mathscr{L}_S(G; \beta_1, \beta_4)$, shown in Fig. 2.11 (far right), the edge e_{13} has weight 17.

Since the edge weights of G are $\omega(E) = \{1, \ldots, 8\}$, these weights belong to the semiring of nonnegative integers $\mathbb{Z}^+ = \{0, 1, 2 \ldots\}$. As guaranteed by Theorem 2.9, the graph $\mathscr{L}_S(G)$ also has edge weights in $\mathbb{Z}^+$. Moreover, one can also compute that $\sigma_0(\mathscr{L}_S(G)) = \sigma_0(G)$ and $\rho(\mathscr{L}_S(G)) = \rho(G)$.

By first expanding a graph and then merging its branches, we will typically change the size of the graph. To determine the size of the resulting graph, we will use the following notation. For a graph $G = (V, E, \omega)$, let $|G| = |V|$, i.e., the number of vertices in V.

Proposition 2.1. *For $G = (V, E, \omega)$, let $S = \{v_1, \ldots, v_m\} \in st_0(G)$. Then*

$$|\mathscr{L}_S(G)| = |S| + \sum_{i=1}^{m}(|\beta^j| - 2),$$

where $|\beta^j| = 2$ if $\beta^j = \emptyset$.

Proof. Let $\mathscr{Y}_S(G) = (\mathscr{V}, \mathscr{U}, \nu)$, where we suppose $\alpha, \beta \in \mathscr{B}_S(\mathscr{Y}_S(G))$ and $\alpha \neq \beta$. Note that the branches of $\mathscr{B}_S(\mathscr{Y}_S(G))$ are pairwise independent and each vertex of $\mathscr{V}$ belongs to a branch of $\mathscr{B}_S(\mathscr{Y}_S(G))$. Hence, $\{\alpha\}_{\text{int}} \cap \{\beta\}_{\text{int}} = \emptyset$ and $\{\beta\}_{\text{int}} \cap S = \emptyset$. Therefore, $\mathscr{V}$ is the disjoint union

$$\mathscr{V} = S \cup \left(\bigcup_{\beta \in \mathscr{B}_S(\mathscr{Y}_S(G))} \{\beta\}_{\text{int}}\right).$$

Recall that the set $\mathscr{U}$, given by (2.12), is the set of interior vertices of the branches $\mathscr{B}_S(\mathscr{Y}_S(G)) - \{\beta^1, \ldots, \beta^m\}$. Thus, the vertex set $\mathscr{V} - \mathscr{U}$ is given by

$$\mathscr{V} - \mathscr{U} = S \cup \left(\bigcup_{i=1}^{m} \{\beta^j\}_{\text{int}}\right),$$

2.6 Isospectral Graph Transformations over Modified Weight Sets

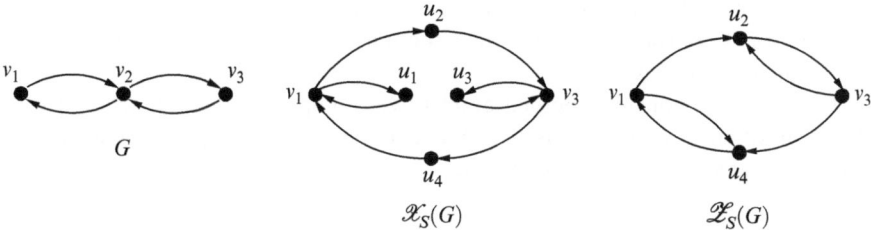

Fig. 2.12 The graphs $\mathscr{X}_S(G)$ and $\mathscr{L}_S(G)$ for $S = \{v_1, v_3\}$, both of which are larger than G

where each $\{\beta^i\}_{\text{int}} \cap \{\beta^j\}_{\text{int}}$ is the empty set and $\{\beta^i\}_{\text{int}} \cap S = \emptyset$ for $i \neq j$. Therefore,

$$|\mathscr{V} - \mathscr{U}| = |S| + \sum_{i=1}^m (|\beta^j| - 2),$$

since each branch β^j has $|\beta^j| - 2$ interior vertices. If $\beta^j = \emptyset$, then this branch has no interior vertices, which is compensated by the assumption $|\beta^j| = 2$.

Note that $|\mathscr{L}'_S S(G)| = |\mathscr{Y}_S(G)|_{\mathscr{V}-\mathscr{U}}| = |\mathscr{V} - \mathscr{U}|$. Since constructing $\mathscr{L}''_S(G)$ from $\mathscr{L}'_G(S)$ involves only the addition of edges, and constructing $\mathscr{L}_S(G)$ from $\mathscr{L}''_G G(S)$ involves only the removal of parallel edges, it follows that $|\mathscr{L}_S(G)| = |\mathscr{V} - \mathscr{U}|$. This completes the proof. □

In Example 2.11, we considered the construction of the graph $\mathscr{L}_S(G)$ in Fig. 2.11 over the complete structural set $S = \{v_1, v_6, v_7\}$. From Proposition 2.1, it follows that $|\mathscr{L}_S(G)| = |S| + |\beta^1| + |\beta^6| + |\beta^7|$. Since $\beta^1 = \emptyset$, $\beta^6 = v_1, v_3, v_6$, and $\beta^7 = v_1, v_2, v_5, v_7$, we have $|\mathscr{L}_S(G)| = 6$.

Note that in this example, $|G| > |\mathscr{L}_S(G)|$. Hence $\mathscr{L}_S(G)$ can be considered a *reduction* of G over the weight set $\mathbb{Z}^+$ having the same nonzero spectrum. However, we note that it may happen that $|\mathscr{L}_\mathscr{S}(\mathscr{G})| > |G|$ for a particular graph G, as shown in the following example.

Example 2.12. Consider the unweighted graph G in Fig. 2.12 (left) with the complete structural set $S = \{v_1, v_3\}$. The graph $\mathscr{L}_S(G)$ shown in Fig. 2.12 (right) is strictly larger than the original graph G, i.e., $\mathscr{L}_S(G)$ has more vertices than G. Therefore, even though the branches of the expansion $\mathscr{X}_S(G)$, shown in Fig. 2.12 (center), have been merged to form $\mathscr{L}_S(G)$, this resulting graph is still larger than the original graph G. At this point, it is yet unknown whether there is a general construction (algorithm) that allows one to reduce an arbitrary graph $G \in \mathbb{G}$ in size while preserving the semiring of its edge weights and (nonzero) spectrum.

As a final remark for this chapter, we note that it is possible to define a merged graph $\mathscr{L}_S(G)$ over a general structural set S rather than a complete structural set. This more general construction is analogous to the procedure described in this section in that $\mathscr{L}_S(G)$ is still constructed by expanding, reweighting, and merging the branches of G. The major difference, though, is that if S is a structural set, then these branches may have loops.

The reason we consider merged graphs over complete structural sets is for the sake of simplicity. We leave it as an exercise to the reader to apply these techniques over structural sets that are not complete.

Chapter 3
Stability of Dynamical Networks

In this chapter, we consider networks from a dynamical systems point of view. When we speak of a system as being "dynamic," what we mean is that the system has a number of possible states and that at each moment of time, the system is in one of those states. The way in which the system's state changes from one moment in time to the next is referred to as the system's *dynamics*.

If a network is dynamic, as most real networks are, the network's *state* at a particular point in time is the collection of each of the states of its elements. That is, the dynamics of a network is the collective dynamics of the network's individual elements.

Network dynamics are a combination of the following two factors. First, every network element may have its own intrinsic dynamics, so that it behaves, in isolation, as its own dynamical system. The second dynamic component of a network comes from the interactions among the various network elements.

Dynamical networks are therefore networks of interacting dynamical systems, which are composed of (i) local dynamical systems, (ii) interactions among these local systems, and (iii) the network's graph of interactions, often called the network topology (see Sect. 2.1). The network's local systems describe the dynamics of the network elements in isolation. The interactions and graph of interactions describe, respectively, the dynamics and structure of the interactions among these elements.

Our major focus in this chapter is the stability of a dynamical network. The particular type of stability we consider relates to whether a network has a globally attracting fixed point.

We analyze the stability of a network in two different settings. In the first, we consider general networks of interacting dynamical systems in which the interactions do not involve time delays. In the second, we allow the interactions to have delays.

We also distinguish two types of delays; those that are single and those that are of a multiple-delay type. By *single*, we mean that the delayed interactions between any two network elements happen on a single time scale. Otherwise, the network has a *multiple-type* delay.

One of the main results of this chapter is that despite the wide variety and potential complexity of such systems, it is possible to give a criterion for the global stability of a general dynamical network. In fact, using this criterion, we prove that a network's stability is invariant with respect to the removal of time delays and the addition of single-type time delays. This leads us to introduce the notion of *intrinsic stability*, which is a stronger form of global stability.

To carry out this analysis, we introduce a new family of graph transformations based on the isospectral graph transformations from Chap. 1. These transformations, which are called *bounded radial transformations* and *isoradial expansions*, are used respectively to modify and maintain the spectral radius of a graph.

We then introduce the notion of an implicit delay in an undelayed dynamical network and show that by removing a network's implicit delays, the result is a lower-dimensional system called a *network restriction*. We then show that such restrictions can be used to obtain improved estimates of a network's global stability.

The reason for this improvement is that network restrictions concentrate the information stored in a network in a smaller set of interactions and elements in the "restricted" network. This consolidation of information leads to improved local estimates and ultimately to an improved estimate of the network's stability.

This idea of consolidating network information to gain improved estimates is an idea used throughout this book. In Chap. 4, this strategy is used to find improved eigenvalue estimates. In Chap. 6, this technique is used to gain improved escape rate estimates in open dynamical systems or, alternatively, rates of information absorption in networks.

The ideas developed in this chapter are illustrated by various examples of Cohen–Grossberg neural networks. We note that this approach to analyzing the global stability of networks is not limited in any way to the study of interacting dynamical systems but can be applied to any time-delayed or undelayed dynamical system.

In particular, this approach is applicable to every multidimensional dynamical system described by a system of simultaneous differential or difference equations. Naturally, it makes sense to apply these techniques only to those systems whose variables depend on one another in some irregular way, e.g., not all variables depend on all other variables.

3.1 Networks as Dynamical Systems

As mentioned in the introduction, dynamical networks are composed of (i) local dynamical systems (ii) interactions between these local systems, and (iii) the network's graph of interactions (see [1, 9, 10]).

To give ourselves a mathematical framework in which to study dynamical networks, we let $\varphi_i : X_i \to X_i$ be maps on the complete metric space (X_i, d) for $i \in \mathscr{I} = \{1, \ldots, n\}$, where

3.1 Networks as Dynamical Systems

$$L_i = \sup_{x_i \neq y_i \in X_i} \frac{d(\varphi_i(x_i), \varphi_i(y_i))}{d(x_i, y_i)} < \infty. \tag{3.1}$$

From the individual spaces (X_i, d), we define $(X, d_{\max})$ to be the metric space on $X = \oplus_{i=1}^n X_i$ with distance given by

$$d_{\max}(\mathbf{x}, \mathbf{y}) = \max_{i \in \mathscr{I}}\{d(x_i, y_i)\}, \; \mathbf{x}, \mathbf{y} \in X.$$

We let (φ, X) be the direct product of the systems (φ_i, X_i) over $i \in \mathscr{I}$, where we refer to (φ, X) as a collection of *local systems*.

Definition 3.1. A map $F : X \to X$ is called an *interaction* if for every $j \in \mathscr{I}$, there exist a nonempty collection of indices $\mathscr{I}_j \subseteq \mathscr{I}$ and a continuous function

$$F_j : \bigoplus_{i \in \mathscr{I}_j} X_i \to X_j,$$

where $F_j(\mathbf{x}) = F_j(\mathbf{x}|_{\mathscr{I}_j})$ for $j \in \mathscr{I}$ and $\mathbf{x} \in X$. The superposition of the interaction and local systems

$$\mathscr{F}(\mathbf{x}) = (F \circ \varphi)(\mathbf{x}) \text{ for } \mathbf{x} \in X$$

generates the dynamical system $(\mathscr{F}, X)$, which is a *dynamical network*.

A given dynamical network $(\mathscr{F}, X)$ may depend on a number of parameters $\mathbf{p} = (p_1, \ldots, p_m)$. If this is the case, we write $\mathscr{F}(\mathbf{x}) = \mathscr{F}(\mathbf{x}; \mathbf{p})$, although we may at times suppress this dependence.

The purpose of this section, beyond introducing the class of dynamical networks we wish to study, is to study the stability of this class of systems. Recall that the dynamical system (f, M) on a metric space M has a *globally attracting fixed point* $\tilde{\mathbf{x}} \in M$ if

$$\lim_{k \to \infty} f^k(\mathbf{x}) = \tilde{\mathbf{x}}, \text{ for any } \mathbf{x} \in M.$$

If a system has a globally attracting fixed point, we call it *globally stable* or simply *stable*.

One of the main questions we wish to address in this chapter is whether for a given collection of local systems (φ, X), a particular interaction leads to a stable dynamical network $(\mathscr{F}, X)$. To consider this, suppose F satisfies the following Lipschitz condition for the finite constants $\Lambda_{ij} \geq 0$:

$$d\left(F_j(\mathbf{x}), F_j(\mathbf{y})\right) \leq \sum_{i \in \mathscr{I}_j} \Lambda_{ij} d(x_i, y_i) \tag{3.2}$$

for all $\mathbf{x}, \mathbf{y} \in X$ and $i, j \in \mathscr{I}$. The Lipschitz constants Λ_{ij} in equation (3.2) form the nonnegative matrix $\Lambda \in \mathbb{R}^{n \times n}$, where $\Lambda_{ij} = 0$ if $i \notin \mathscr{I}_j$.

Using the matrix Λ and the constants L_i given in equation (3.1), we define the matrix

$$\Lambda^T \cdot diag[L_1, \ldots, L_n] = \begin{pmatrix} \Lambda_{11} L_1 & \cdots & \Lambda_{n1} L_n \\ \vdots & \ddots & \vdots \\ \Lambda_{1n} L_1 & \cdots & \Lambda_{nn} L_n \end{pmatrix} \in \mathbb{R}^{n \times n},$$

which we refer to as a *stability matrix* of $(\mathscr{F}, X)$. This notion of a stability matrix is used in the following theorem to give a sufficient condition for the stability of the dynamical network $(\mathscr{F}, X)$.

Theorem 3.1. *Suppose A is a stability matrix of the dynamical network $(\mathscr{F}, X)$. If the spectral radius $\rho(A)$ is less than 1, then the dynamical network $(\mathscr{F}, X)$ is stable.*

Before proving Theorem 3.1, we observe that if $\mathbf{v} \in \mathbb{C}^{n \times 1}$, then its ℓ^∞ norm is $||\mathbf{v}||_\infty = \max_i |v_i|$. Moreover, the ℓ^∞ norm of a matrix $A \in \mathbb{C}^{n \times n}$ is

$$|||A|||_\infty = \max_{1 \leq i \leq n} \sum_{j=1}^{n} |a_{ij}|,$$

which is the maximum absolute row sum of A. The ℓ^∞ matrix norm has the additional property that it is *submultiplicative*, i.e., $|||AB|||_\infty \leq |||A|||_\infty |||B|||_\infty$ for every $A, B \in \mathbb{C}^{n \times n}$. With this in place, we give a proof of Theorem 3.1.

Proof. For $\mathbf{x}, \mathbf{y} \in X$ and $1 \leq j \leq n$,

$$\begin{aligned} d\left(\mathscr{F}(\mathbf{x})_j, \mathscr{F}(\mathbf{y})_j\right) &= d\left(F_j(\varphi(\mathbf{x})), F_j(\varphi(\mathbf{y}))\right) \\ &\leq \sum_{i \in \mathscr{I}} \Lambda_{ij} d\left(\varphi_i(x_i), \varphi_i(y_i)\right) \\ &\leq \sum_{i \in \mathscr{I}} \Lambda_{ij} L_i d(x_i, y_i). \end{aligned}$$

Therefore, each entry of the column vector

$$\begin{bmatrix} d\left(\mathscr{F}_1(\mathbf{x}), \mathscr{F}_1(\mathbf{y})\right) \\ \vdots \\ d\left(\mathscr{F}_n(\mathbf{x}), \mathscr{F}_n(\mathbf{y})\right) \end{bmatrix} \leq A \begin{bmatrix} d(x_1, y_1) \\ \vdots \\ d(x_n, y_n) \end{bmatrix}.$$

3.1 Networks as Dynamical Systems

Since

$$\begin{bmatrix} d(\mathscr{F}_1^2(\mathbf{x}), \mathscr{F}_1^2(\mathbf{y})) \\ \vdots \\ d(\mathscr{F}_n^2(\mathbf{x}), \mathscr{F}_n^2(\mathbf{y})) \end{bmatrix} \leq A \begin{bmatrix} d(\mathscr{F}_1(\mathbf{x}), \mathscr{F}_1(\mathbf{y})) \\ \vdots \\ d(\mathscr{F}_n(\mathbf{x}), \mathscr{F}_n(\mathbf{y})) \end{bmatrix} \leq A^2 \begin{bmatrix} d(x_1, y_1) \\ \vdots \\ d(x_n, y_n) \end{bmatrix}$$

by the same reasoning, it follows inductively that

$$d_{\max}(\mathscr{F}^k(\mathbf{x}), \mathscr{F}^k(\mathbf{y})) \leq \left\| A^k \begin{bmatrix} d(x_1, y_1) \\ \vdots \\ d(x_n, y_n) \end{bmatrix} \right\|_\infty \quad (3.3)$$

for all $k > 0$.

By the *Jordan canonical form theorem*, there exist a nonsingular matrix $B \in \mathbb{C}^{n \times n}$ and a block-diagonal matrix $J \in \mathbb{C}^{n \times n}$ such that $A = BJB^{-1}$. In particular,

$$J = \begin{bmatrix} J_{m_1}(\lambda_1) & 0 & \cdots & 0 \\ 0 & J_{m_2}(\lambda_2) & & \vdots \\ \vdots & & \ddots & 0 \\ 0 & \cdots & 0 & J_{m_t}(\lambda_t) \end{bmatrix},$$

where

$$J_{m_i}(\lambda_i) = \begin{bmatrix} \lambda_i & 1 & 0 & \cdots & 0 \\ 0 & \lambda_i & 1 & & 0 \\ \vdots & & \ddots & \ddots & \vdots \\ 0 & & & \lambda_i & 1 \\ 0 & 0 & \cdots & 0 & \lambda_i \end{bmatrix} \in \mathbb{C}^{m_i \times m_i}$$

and $\lambda_i \in \sigma(A)$ for $1 \leq i \leq t$.

Moreover, since J is block diagonal, it follows that for $k \geq 0$,

$$J^k = \begin{bmatrix} J_{m_1}^k(\lambda_1) & 0 & \cdots & 0 \\ 0 & J_{m_2}^k(\lambda_2) & & \vdots \\ \vdots & & \ddots & 0 \\ 0 & \cdots & 0 & J_{m_t}^k(\lambda_t) \end{bmatrix}.$$

Since the ℓ^∞ norm of a matrix is its maximum absolute row sum, we have

$$|||J^k|||_\infty = \max_{1 \leq i \leq t} |||J_{m_i}^k(\lambda_i)|||_\infty.$$

The kth power of a Jordan block $J_{m_i}(\lambda_i)$ can be computed to be

$$J_{m_i}^k(\lambda_i) = \begin{bmatrix} \binom{k}{0}\lambda_i^k & \binom{k}{1}\lambda_i^{k-1} & \binom{k}{2}\lambda_i^{k-2} & \cdots & \binom{k}{m_i-1}\lambda_i^{k-m_i+1} \\ 0 & \binom{k}{0}\lambda_i^k & \binom{k}{1}\lambda_i^{k-1} & \cdots & \binom{k}{m_i-2}\lambda_i^{k-m_i+2} \\ \vdots & & \ddots & & \vdots \\ 0 & & & \binom{k}{0}\lambda_i^k & \binom{k}{1}\lambda_i^{k-1} \\ 0 & 0 & \cdots & 0 & \binom{k}{0}\lambda_i^k \end{bmatrix}$$

for every $k \geq m_i - 1$. Hence for $k \geq m_i - 1$, the matrix norm is given by

$$|||J_{m_i}^k(\lambda_i)|||_\infty = \sum_{j=0}^{m_i-1} \left|\binom{k}{j}\lambda_i^{k-j}\right|$$

$$= \left[\left|\binom{k}{0}\lambda_i^{m_i-1}\right| + \left|\binom{k}{1}\lambda_i^{m_i-2}\right| + \cdots + \left|\binom{k}{m_i-1}\right|\right]|\lambda_i^{k-m_i+1}|$$

$$\leq C_{m_i}\rho(A)^{k-m_i+1},$$

where $C_{m_i} = \sum_{j=0}^{m_i-1} \left|\binom{k}{j}\lambda_i^{m_i-j-1}\right|$.

Suppose $\rho(A) < 1$. Letting $m = \max_{0 \leq i \leq t} m_i$ and $C = \max_{1 \leq i \leq t} C_{m_i}$ yields

$$|||J^k|||_\infty \leq C\rho(A)^{k-m+1} \text{ for all } k \geq m-1.$$

Via equation (3.3), we obtain

$$\sum_{k=m}^\infty d_{\max}(\mathscr{F}^k(\mathbf{x}), \mathscr{F}^k(\mathbf{y})) \leq \sum_{k=m}^\infty \left\|A^k \begin{bmatrix} d(x_1, y_1) \\ \vdots \\ d(x_n, y_n) \end{bmatrix}\right\|_\infty$$

$$\leq \sum_{k=m}^\infty |||BJ^kB^{-1}|||_\infty \left\|\begin{bmatrix} d(x_1, y_1) \\ \vdots \\ d(x_n, y_n) \end{bmatrix}\right\|_\infty$$

$$\leq |||B|||_\infty |||B^{-1}|||_\infty \left\|\begin{bmatrix} d(x_1, y_1) \\ \vdots \\ d(x_n, y_n) \end{bmatrix}\right\|_\infty \sum_{k=m}^\infty |||J^k|||_\infty$$

$$\leq |||B|||_\infty |||B^{-1}|||_\infty \left\|\begin{bmatrix} d(x_1, y_1) \\ \vdots \\ d(x_n, y_n) \end{bmatrix}\right\|_\infty \frac{C\rho(A)}{1-\rho(A)} < \infty.$$

Letting $\mathbf{y} = \mathscr{F}(\mathbf{x})$, it then follows that $\sum_{k=m}^\infty d_{\max}(\mathscr{F}^k(\mathbf{x}), \mathscr{F}^{k+1}(\mathbf{x})) < \infty$.

Hence, for $\epsilon > 0$, there exists N such that $\sum_{k=N}^{\infty} d_{\max}\left(\mathscr{F}^k(\mathbf{x}), \mathscr{F}^{k+1}(\mathbf{x})\right) < \epsilon$. Therefore,

$$d_{\max}\left(\mathscr{F}^k(\mathbf{x}), \mathscr{F}^\ell(\mathbf{x})\right) < \epsilon$$

for every $k, \ell > N$, implying that the sequence $\{F^k(\mathbf{x})\}_{k \geq 1}$ is Cauchy. Since X is complete, this sequence converges.

If $\mathscr{F}^k(\mathbf{x}) \to \tilde{\mathbf{x}}$, then since $\mathscr{F}$ is continuous, it follows that $\mathscr{F}(\tilde{\mathbf{x}}) = \tilde{\mathbf{x}}$. Moreover, since $\sum_{k=1}^{\infty} d_{\max}\left(\mathscr{F}^k(\tilde{\mathbf{x}}), \mathscr{F}^k(\mathbf{y})\right) \leq \infty$ for every $\mathbf{y} \in X$, it follows that $\mathscr{F}^k(\mathbf{y}) \to \tilde{\mathbf{x}}$, which completes the proof. □

We note that the condition for stability given in Theorem 3.1 is a sufficient condition. Indeed, it is possible to find systems that are stable but do not have a stability matrix A such that $\rho(A) < 1$. A natural question, then, is what it means for a dynamical network to be stable in the sense of Theorem 3.1.

As we will demonstrate, the stability condition in Theorem 3.1 guarantees more than that a dynamical network is stable. For this reason, we give this type of stability the following name.

Definition 3.2. If $\rho(A) < 1$, where A is a stability matrix of $(\mathscr{F}, X)$, then we say that this network is *intrinsically stable*.

For now, Theorem 3.1 implies that an intrinsically stable network is globally stable. Later, in Sect. 3.2, it will be shown that if a network is intrinsically stable, then it remains stable even if certain types of time delays are added to its dynamics (see Theorem 3.4). Since the addition of time delays can destabilize a network, intrinsic stability is in this sense a stronger form of global stability.

Before continuing, we note that a dynamical network $(\mathscr{F}, X)$ need not be differentiable to have a stability matrix. However, if a network is differentiable, then not only is it computationally easier to compute a stability matrix of the network, it is possible to compute the network's optimal stability matrix.

To demonstrate this, suppose the maps $\varphi : X \to X$ and $F : X \to X$ are continuously differentiable and each $X_i \subseteq \mathbb{R}$ is a closed but possibly unbounded interval. If the constants

$$L_i = \max_{\mathbf{x} \in X} |\varphi_i'(x_i)| < \infty, \qquad (3.4)$$

$$\Lambda_{ij} = \max_{\mathbf{x} \in X} |(DF)_{ji}(\mathbf{x})| < \infty, \qquad (3.5)$$

where DF is the matrix of first partial derivatives of F, then we write $\mathscr{F} \in C_\infty^1(X)$. For $\mathscr{F} \in C_\infty^1(X)$, the matrix

$$A_{\mathscr{F}} = \Lambda^T \cdot diag[L_1, \ldots, L_n]$$

given by (3.4) and (3.5) can be shown to be a stability matrix of $(\mathscr{F}, X)$.

The matrix $A_\mathscr{F}$ can be shown to be optimal for determining whether $(\mathscr{F}, X)$ is intrinsically stable. However, to demonstrate this, we let $\rho(\mathscr{F}) = \rho(A_\mathscr{F})$ denote the *spectral radius* of $(\mathscr{F}, X)$ and note the following. For the matrices $A, B \in \mathbb{R}^{n \times n}$, we write $A \leq B$ if $A_{ij} \leq B_{ij}$ for each $1 \leq i, j \leq n$. If it is known that $0 \leq A \leq B$, where 0 is the zero matrix, then $\rho(A) \leq \rho(B)$ (see, for instance, [22]). This is used to prove the following.

Proposition 3.1. *Suppose* $\mathscr{F} \in C^1_\infty(X)$. *If B is any stability matrix of* $(\mathscr{F}, X)$, *then* $\rho(\mathscr{F}) \leq \rho(B)$.

Proof. Suppose $\mathscr{F} \in C^1_\infty(X)$ and that $B = \tilde{\Lambda}^T \cdot diag[\tilde{L}_1, \ldots, \tilde{L}_n]$ is a stability matrix of $(\mathscr{F}, X)$. For $\mathbf{e}_i$, the ith standard basis vector of $\mathbb{R}^n$, and $\mathbf{x} \in X$ let $h \neq 0$ such that $\mathbf{y} = \mathbf{x} + h\mathbf{e}_i \in X$. Then by (3.2),

$$d\big(F_j(\mathbf{x}), F_j(\mathbf{y})\big) \leq \tilde{\Lambda}_{ij} d(x_i, x_i + h) = \tilde{\Lambda}_{ij}|h|.$$

Hence $|F(\mathbf{x})_j - F(\mathbf{y})_j|/|h| \leq \tilde{\Lambda}_{ij}$. Taking the limit as $h \to 0$, we have

$$|DF_{ji}(\mathbf{x})| \leq \tilde{\Lambda}_{ij} \text{ for all } \mathbf{x} \in X.$$

By setting $\Lambda_{ij} = \max_{\mathbf{x} \in X} |DF_{ji}(\mathbf{x})|$, it follows that $0 \leq \Lambda \leq \tilde{\Lambda}$.

Similarly, one can show that $L_i = \max_{\mathbf{x} \in X} |\varphi'_i(x_i)| \leq \tilde{L}_i$. Hence the matrix $A_\mathscr{F}$ is less than or equal to B implying $\rho(\mathscr{F}) \leq \rho(B)$. $\square$

For $\mathscr{F} \in C^1_\infty(X)$, Proposition 3.1 implies that $\rho(\mathscr{F})$ is optimal for directly determining, via Theorem 3.1, whether $(\mathscr{F}, X)$ is intrinsically stable.

Remark 3.1. In what follows, we formally consider those dynamical networks $(\mathscr{F}, X)$ for which $\mathscr{F} \in C^1_\infty(X)$. We note that the results in this chapter can be shown to hold for every network satisfying (3.1) and (3.2) by slight modifications of the proofs we give. The reason we restrict our discussion to the class $C^1_\infty(X)$ is to simplify the exposition.

As an example of the usefulness of Theorem 3.1, let $(\varphi_i, \mathbb{R})$ be the local systems

$$\varphi_i(x_i) = (1 - \epsilon)x_i + c_i, \tag{3.6}$$

where $c_i, \epsilon \in \mathbb{R}$ and $1 \leq i \leq n$. We note that these systems are stable if and only if $|1 - \epsilon| < 1$. We then ask what kind of interaction between these local systems will lead to a stable dynamical network.

For the local systems given by (3.6), we are specifically interested in interactions of the form

$$C_j(\mathbf{x}) = x_j + \sum_{j=1}^n W_{ij}\phi_i\left(\frac{x_i - c_i}{1 - \epsilon}\right), \tag{3.7}$$

3.1 Networks as Dynamical Systems

where $\epsilon \neq 1$, $W \in \mathbb{R}^{n \times n}$, and $\phi_i : \mathbb{R} \to \mathbb{R}$ is any smooth sigmoidal function with Lipschitz constant $\mathscr{L} \geq 0$. That is, ϕ_i is a bounded differentiable function such that $\phi_i'(x) > 0$ for all $x \in \mathbb{R}$.

The reason we consider this particular interaction is that the dynamical network $(\mathscr{C}, \mathbb{R}^n)$ with $\mathscr{C} = C \circ \varphi$ is then given by

$$\mathscr{C}_j(\mathbf{x}) = (1 - \epsilon)x_j + \sum_{i=1}^{n} W_{ij} \phi(x_i) + c_j, \tag{3.8}$$

which is a special case of a Cohen–Grossberg neural network in discrete time [17]. Because of the large number of parameters involved, we suppress the dependence of the function $\mathscr{C}(\mathbf{x})$ on $c_1, \ldots, c_n$, ϵ, and W in those systems that resemble Cohen–Grossberg neural networks.

For such neural networks, the variable x_i represents the *activation* of the ith neuron population. Here, the function ϕ_i describes the ith neuron population's response to inputs. The matrix W gives the interaction strengths between each of the ith and jth neuron populations, which describe how the neurons are connected within the network. The constants c_i indicate constant inputs from outside the system.

To determine a stability criterion for the dynamical network $(\mathscr{C}, \mathbb{R}^{n \times n})$, we let $|W|$ denote the matrix with entries $|W|_{ij} = |W_{ij}|$. The following result gives a general stability condition for this class of Cohen–Grossberg neural networks.

Theorem 3.2 (Stability of Cohen–Grossberg Neural Networks). *Let $(\mathscr{C}, \mathbb{R}^n)$ be the Cohen–Grossberg network given by (3.8), where ϕ_i has Lipschitz constant $\mathscr{L}$. If*

$$|1 - \epsilon| + \mathscr{L}\rho(|W|) < 1,$$

then $(\mathscr{C}, \mathbb{R}^n)$ is intrinsically stable.

Proof. The claim is that for the local systems and interaction given by (3.6) and (3.7), the matrix $A = \tilde{\Lambda}^T \cdot diag[L_1, \ldots, L_n]$ with

$$\max_{\mathbf{x} \in X} |\varphi_i'(x_i)| = L_i = |1 - \epsilon|, \text{ and}$$

$$\max_{\mathbf{x} \in X} |(DC)_{ji}(\mathbf{x})| \leq \tilde{\Lambda}_{ij} = \begin{cases} \left| 1 + \frac{W_{ji}\mathscr{L}}{1-\epsilon} \right| & \text{for } i = j, \\ \left| \frac{W_{ji}\mathscr{L}}{1-\epsilon} \right| & \text{for } i \neq j \end{cases}$$

is a stability matrix of $(\mathscr{C}, \mathbb{R}^n)$.

To see this, note that the constants

$$\Lambda_{ij} = \max_{\mathbf{x} \in X} |(DC)_{ji}(\mathbf{x})|$$

satisfy (3.2). Since $\Lambda_{ij} \leq \tilde{\Lambda}_{ij}$, the constants $\tilde{\Lambda}_{ij}$ must also satisfy (3.2), which verifies the claim.

Since the matrix A has the form

$$A = \begin{bmatrix} |1-\epsilon| + |W_{11}\mathscr{L}| & |W_{12}\mathscr{L}| & \cdots & |W_{1n}\mathscr{L}| \\ |W_{21}\mathscr{L}| & |1-\epsilon| + |W_{22}\mathscr{L}| & \cdots & |W_{2n}\mathscr{L}| \\ \vdots & \vdots & \ddots & \vdots \\ |W_{n1}\mathscr{L}| & |W_{n2}\mathscr{L}| & \cdots & |1-\epsilon| + |W_{nn}\mathscr{L}| \end{bmatrix},$$

the spectral radius of the matrix A is then

$$\rho(A) = \rho\big(|1-\epsilon|I + \mathscr{L}|W|\big) = |1-\epsilon| + \mathscr{L} \cdot \rho(|W|).$$

Therefore, if $|1-\epsilon| + \mathscr{L}\rho(|W|) < 1$, then $(\mathscr{C}, \mathbb{R}^n)$ is intrinsically stable. □

One of the major goals of this chapter is to extend results, such as Theorems 3.1 and 3.2, to the case in which the network's interactions include time delays. However, this requires that we first consider the case in which a dynamical network has trivial local dynamics.

For a given dynamical network, it is always formally possible to absorb the dynamics of the network's local systems into its interaction. This is done by considering the dynamical network $(\mathscr{F}, X)$ to have the interaction $F \circ \varphi$ and no local dynamics, i.e., local dynamics given by the identity map $Id : X \to X$.

In this sense, the theory we present here can be used to deal with general dynamical systems, which typically do not have local dynamics. That is, in many systems, the dynamics of each component in the absence of the others cannot be explicitly defined.

If we have both a collection of local systems (φ, X) and an interaction (F, X), the network $(\mathscr{F}, X)$ can always be considered to be a network with no local dynamics. If only the local systems (φ, X) are given, the question is then what kind of interaction induces a specific type of network dynamics. For instance, what type of interaction induces stability?

The difference between these points of view is that the first focuses on the network interactions, while the second emphasizes local dynamics. Both are relevant, but for the moment, we consider the case in which we have a fixed interaction, which will be important in the following sections.

By absorbing the network's local dynamics into its interaction, we can consider $(\mathscr{F}, X)$ to be the dynamical network with interaction $F \circ \varphi$ and no local dynamics, i.e., local dynamics given by $Id : X \to X$. By way of notation, we let $(\underline{\mathscr{F}}, X)$ denote the network $(\mathscr{F}, X)$ considered as a network without local dynamics.

Proposition 3.2. *Assuming $\mathscr{F} \in C^1_\infty(X)$, then $\rho(\underline{\mathscr{F}}) \leq \rho(\mathscr{F})$, where $A_{\underline{\mathscr{F}}}$ is given by*

$$(A_{\underline{\mathscr{F}}})_{ij} = \max_{x \in X} |D\mathscr{F}_{ji}(x)|. \tag{3.9}$$

Proof. Since $(\tilde{\mathcal{F}}, X)$ has no local dynamics and the continuously differentiable interaction $\tilde{\mathcal{F}}$ is equal to $F \circ \varphi$, it follows that $L_i = 1$ satisfies (3.4), and $\max_{\mathbf{x} \in X} |D\tilde{\mathcal{F}}_{ji}(\mathbf{x})|$ satisfies (3.5). Since $diag[L_1, \ldots, L_n] = I$, the $n \times n$ identity matrix, it follows that equation (3.9) holds.

To show that $\rho(\tilde{\mathcal{F}}) \leq \rho(\mathcal{F})$, note that

$$(A_{\tilde{\mathcal{F}}})_{ij} = \max_{\mathbf{x} \in X} |D\tilde{\mathcal{F}}_{ji}(\mathbf{x})| = \max_{\mathbf{x} \in X} \left| \frac{\partial F_j}{\partial x_i}(\mathbf{x}) \varphi_i'(x_i) \right| \leq$$

$$\max_{\mathbf{x} \in X} \left| \frac{\partial F_j}{\partial x_i}(\mathbf{x}) \right| \max_{x_i \in X_i} |\varphi_i'(x_i)| = \Lambda_{ij} L_i.$$

Hence, $A_{\tilde{\mathcal{F}}} \leq \Lambda^T \cdot diag[L_1, \ldots, L_n]$, implying $\rho(A_{\tilde{\mathcal{F}}}) \leq \rho(\Lambda^T \cdot diag[L_1, \ldots, L_n])$, and the result follows. □

Proposition 3.2 has the following consequence. Since $\rho(\tilde{\mathcal{F}}) \leq \rho(\mathcal{F})$, then by considering $(\tilde{\mathcal{F}}, X)$ rather than $(\mathcal{F}, X)$, we obtain an improved estimate of the network's stability, since these systems have equivalent dynamics. The reason for this improved estimate is that in $(\tilde{\mathcal{F}}, X)$, we have integrated different parts of the network's dynamics, i.e., the local dynamics and interaction are combined.

This technique of compressing network information is the basic method used later, in Sect. 3.4, to get improved stability estimates of $(\mathcal{F}, X)$ by removing the network's implicit time delays. However, before considering such improvements, we first turn our attention to analyzing the stability of time-delayed networks.

3.2 Time-Delayed Dynamical Networks

In most real networks, the network elements are physically separated by some distance. Additionally, these elements are often used to process incoming information. The time required to send signals over a distance and to process that information inevitably leads to time delays in the network's interaction. These time delays are important to the network's dynamics, since they are often a source of instability and a cause of poor performance.

Our first task in this section is to extend our mathematical framework to include dynamical networks with time delays. This requires that we allow the state $\mathcal{F}^{k+1}(\mathbf{x})$ of the dynamical network $(\mathcal{F}, X)$ to depend not only on $\mathcal{F}^k(\mathbf{x})$ but on some subset of the previous T states $\mathcal{F}^k(\mathbf{x}), \mathcal{F}^{k-1}(\mathbf{x}), \ldots, \mathcal{F}^{k-(T-1)}(\mathbf{x})$ of the system.

Given the fixed integer $T \geq 1$, let $\mathcal{T} = \{0, -1, \ldots, -T+1\}$. We define the product space

$$X^T = \bigoplus_{\tau \in \mathcal{T}} X,$$

where a point $\mathbf{x} = (\mathbf{x}^0, \ldots, \mathbf{x}^{-T+1}) \in X^T$ has coordinates $(\mathbf{x})^\tau = \mathbf{x}^\tau \in X$ and where $(\mathbf{x})^\tau_i = x^\tau_i \in X_i$ for $(i, \tau) \in \mathscr{I} \times \mathscr{T}$. We say that $\mathbf{x}^\tau$ is $\mathbf{x}$ at time τ and that x^τ_i is the i th component of $\mathbf{x}^\tau$.

Given the local systems (φ, X), we define the function $\bar{\varphi} : X^T \to X^T$ as follows. For $\mathbf{x} = (\mathbf{x}^0, \ldots, \mathbf{x}^{-T+1}) \in X^T$, let

$$\bar{\varphi}(\mathbf{x}^0, \mathbf{x}^{-1}, \ldots, \mathbf{x}^{-T+1}) = \left(\varphi(\mathbf{x}^0), \mathbf{x}^{-1}, \ldots, \mathbf{x}^{-T+1}\right),$$

so that only the *present* ($\tau = 0$) time step of $\mathbf{x}$ is affected by φ.

Definition 3.3. A map $H : X^T \to X$ is called a *time-delayed interaction* if for all $j \in \mathscr{I}$, there exist a nonempty set $\mathscr{I}^j \subseteq \mathscr{I} \times \mathscr{T}$ and a function

$$H_j : \bigoplus_{(i,\tau) \in \mathscr{I}^j} X^\tau_i \to X_j.$$

The map H is defined by $H_j(\mathbf{x}) = H_j(\mathbf{x}|_{\mathscr{I}^j})$ for $j \in \mathscr{I}$ and $\mathbf{x} \in X^T$. The superposition of the local systems and time-delayed interaction

$$\mathscr{H}(\mathbf{x}) = (H \circ \bar{\varphi})(\mathbf{x}) \text{ for } \mathbf{x} \in X^T$$

generates the *time-delayed dynamical network* $(\mathscr{H}, X^T)$. The *orbit* of the point $(\mathbf{x}^0, \mathbf{x}^{-1}, \ldots, \mathbf{x}^{-T+1}) \in X^T$ under $\mathscr{H}$ is the sequence $\{\mathbf{x}^k\}_{k > -T}$, where

$$\mathbf{x}^{k+1} = \mathscr{H}(\mathbf{x}^k, \mathbf{x}^{k-1}, \ldots, \mathbf{x}^{k-T+1}).$$

Here we assume that both φ and H are continuously differentiable, so that by a slight abuse of our notation, $\mathscr{H} \in C^1_\infty(X^T)$. As in Sect. 3.1, this assumption is for convenience. We note that the results in this section can be extended to those time-delayed dynamical networks $(\mathscr{H}, X^T)$ that are Lipschitz continuous on the complete metric space X^T (see Remark 3.1).

Remark 3.2. For the sake of generality, we could assume that the local systems (φ, X^T) are also time delayed. However, if such is the case, we can absorb these local time delays into the delayed interaction. Thus, the mathematical framework we present here can be used to investigate the more general case of interacting systems that have time delays in both the local systems and interaction.

In this section, we will again concentrate on finding a sufficient condition under which a time-delayed dynamical network has a globally attracting fixed point. A *fixed point* of $(\mathscr{H}, X^T)$ is an $\tilde{\mathbf{x}} \in X$ such that $\tilde{\mathbf{x}} = \mathscr{H}(\tilde{\mathbf{x}}, \ldots, \tilde{\mathbf{x}})$. The fixed point $\tilde{\mathbf{x}} \in X$ is a *global attractor* of $(\mathscr{H}, X^T)$ if for every initial condition $(\mathbf{x}^0, \mathbf{x}^{-1}, \ldots, \mathbf{x}^{-T+1}) \in X^T$, we have the limit

$$\lim_{k \to \infty} \mathbf{x}^k = \tilde{\mathbf{x}}.$$

As before, we call the time-delayed dynamical network $(\mathscr{H}, X^T)$ *stable* if it has a globally attracting fixed point.

3.2 Time-Delayed Dynamical Networks

Since time-delayed networks are similar in many ways to undelayed networks, it is tempting to use the results given in Sect. 3.1 to determine whether $(\mathcal{H}, X^T)$ is stable. However, in order to do so, we must first modify the function $\mathcal{H}$.

Definition 3.4. For the time-delayed $(\mathcal{H}, X^T)$, let $\bar{\mathcal{H}} : X^T \to X^T$ be the map

$$\bar{\mathcal{H}}(\mathbf{x}) = (\mathcal{H}(\mathbf{x}), \mathbf{x}^0, \mathbf{x}^{-1}, \ldots, \mathbf{x}^{-T+2})$$

for $\mathbf{x} = (\mathbf{x}^0, \mathbf{x}^{-1}, \ldots, \mathbf{x}^{-T+1}) \in X^T$. We call $(\bar{\mathcal{H}}, X^T)$ the *dynamical network* associated with $(\mathcal{H}, X^T)$.

We assume that the dynamical network $(\bar{\mathcal{H}}, X^T)$ has no local dynamics. That is, we let $\bar{\mathcal{H}}(\mathbf{x}) = (\bar{\mathcal{H}} \circ Id)(\mathbf{x})$, so that the network has the interaction $\bar{\mathcal{H}} : X^T \to X^T$ and local dynamics $Id : X^T \to X^T$.

Remark 3.3. If the local systems (φ, X) are invertible, then it is possible to define $(\bar{\mathcal{H}}, X^T)$ as a dynamical network with the nontrivial local systems $(\bar{\varphi}, X^T)$ and interaction

$$\bar{H}(\mathbf{x}) = (H(\mathbf{x}), \varphi^{-1}(\mathbf{x}^0), \mathbf{x}^{-1}, \ldots, \mathbf{x}^{-T+2}).$$

The dynamical network $(\bar{\mathcal{H}}, X^T)$ is assumed, for the sake of generality, to have the interaction $\bar{\mathcal{H}}$ and no local dynamics.

The dynamical network $(\bar{\mathcal{H}}, X^T)$ has components

$$\bar{\mathcal{H}}_i^\tau(\mathbf{x}) = \begin{cases} \mathcal{H}_i(\mathbf{x}) & \tau = 0, \\ x_i^{\tau+1} & -T < \tau < 0 \end{cases} \quad \text{for } i \in \mathcal{I}. \tag{3.10}$$

This, together with (3.9), implies that the stability matrix $A_{\bar{\mathcal{H}}}$ has the entries

$$(A_{\bar{\mathcal{H}}})_{ij} = \max_{\mathbf{x} \in X^T} |(D\bar{\mathcal{H}})_{ji}(\mathbf{x})| \tag{3.11}$$

for $i, j \in \mathcal{I} \times \mathcal{T}$. Additionally, since $(\bar{\mathcal{H}}^k(\mathbf{x}))^0 = \mathcal{H}(\mathbf{x}^k, \mathbf{x}^{k-1}, \ldots, \mathbf{x}^{-T+1})$ for every initial condition $\mathbf{x} = (\mathbf{x}^0, \mathbf{x}^{-1}, \ldots, \mathbf{x}^{-T+1}) \in X^T$ and $k \geq 0$, the following holds.

Lemma 3.1. *The time-delayed network $(\mathcal{H}, X^T)$ is stable if and only if its associated dynamical network $(\bar{\mathcal{H}}, X^T)$ is stable.*

In light of lemma 3.1, we let $\rho(\mathcal{H}) = \rho(\bar{\mathcal{H}})$ denote the spectral radius of $(\mathcal{H}, X^T)$ and say that $(\mathcal{H}, X^T)$ is *intrinsically stable* if $\rho(\mathcal{H}) < 1$. By combining this result with Theorem 3.1, it is possible to investigate the dynamic stability of $(\mathcal{H}, X^T)$ via the dynamical network $(\bar{\mathcal{H}}, X^T)$.

Corollary 3.1. *If $\rho(\mathcal{H}) < 1$, then the time-delayed dynamical network $(\mathcal{H}, X^T)$ is stable.*

By associating a time-delayed network with a network that formally does not have delays, it is possible to use the theory developed in Sect. 3.1 to study the stability of time-delayed dynamical networks. This approach is illustrated in the following example.

Example 3.1. Consider the time-delayed dynamical network $(\mathcal{H}, X^T)$ given by

$$\mathcal{H}(\mathbf{x}^k, \mathbf{x}^{k-1}, \mathbf{x}^{k-2}, \mathbf{x}^{k-3}) = \begin{bmatrix} (1-\epsilon)x_1^{k-1} + 2a\tanh(bx_2^{k-3}) + c_1 \\ (1-\epsilon)x_2^{k-1} + 2a\tanh(bx_1^{k-3}) + c_2 \end{bmatrix} \quad (3.12)$$

with local systems $\varphi_i(x_i) = (1-\epsilon)x_i + c_i$ for $i = 1, 2$ and interaction

$$H(\mathbf{x}^k, \mathbf{x}^{k-1}, \mathbf{x}^{k-2}, \mathbf{x}^{k-3}) = \begin{bmatrix} x_1^{k-1} + 2a\tanh(b\frac{x_2^{k-3}-c_2}{1-\epsilon}) \\ x_2^{k-1} + 2a\tanh(b\frac{x_1^{k-3}-c_1}{1-\epsilon}) \end{bmatrix}, \quad (3.13)$$

where $a, b, c_i \in \mathbb{R}$, $X = \mathbb{R}^2$, $T = 4$, and $\epsilon \neq 1$.

Note that this system has the form of a Cohen–Grossberg network, considered in Sect. 3.1, but with time delays. The function $\phi_i(x_i) = \tanh(\mathcal{L}x_i)$ is a standard sigmoidal function considered in the theory of Cohen–Grossberg neural networks (see, for example, [29]).

The dynamical network $(\bar{\mathcal{H}}, X^T)$ associated with this system has components

$$\bar{\mathcal{H}}(\mathbf{x}) = \begin{bmatrix} \mathcal{H}_1^0(x_1^{-1}, x_2^{-3}) \\ \mathcal{H}_1^{-1}(x_1^0) \\ \mathcal{H}_1^{-2}(x_1^{-1}) \\ \mathcal{H}_1^{-3}(x_1^{-2}) \\ \mathcal{H}_2^0(x_1^{-3}, x_2^{-1}) \\ \mathcal{H}_2^{-1}(x_2^0) \\ \mathcal{H}_2^{-2}(x_2^{-1}) \\ \mathcal{H}_2^{-3}(x_2^{-2}) \end{bmatrix} = \begin{bmatrix} (1-\epsilon)x_1^{-1} + 2a\tanh(bx_2^{-3}) + c_1 \\ x_1^0 \\ x_1^{-1} \\ x_1^{-2} \\ (1-\epsilon)x_2^{-1} + 2a\tanh(bx_1^{-3}) + c_2 \\ x_2^0 \\ x_2^{-1} \\ x_2^{-2} \end{bmatrix}.$$

Using equation (3.11), we compute the stability matrix $A_{\bar{\mathcal{H}}}$ to be

$$A_{\bar{\mathcal{H}}} = \begin{bmatrix} 0 & 1 & 0 & 0 & 0 & 0 & 0 & 0 \\ |1-\epsilon| & 0 & 1 & 0 & 0 & 0 & 0 & 0 \\ 0 & 0 & 0 & 1 & 0 & 0 & 0 & 0 \\ 0 & 0 & 0 & 0 & |2ab| & 0 & 0 & 0 \\ 0 & 0 & 0 & 0 & 0 & 1 & 0 & 0 \\ 0 & 0 & 0 & 0 & |1-\epsilon| & 0 & 1 & 0 \\ 0 & 0 & 0 & 0 & 0 & 0 & 0 & 1 \\ |2ab| & 0 & 0 & 0 & 0 & 0 & 0 & 0 \end{bmatrix}.$$

3.2 Time-Delayed Dynamical Networks

Taking the spectral radius of this matrix, we find that

$$\rho(A_{\bar{\mathcal{H}}}) = \sqrt{\frac{|1-\epsilon| + \sqrt{|1-\epsilon|^2 + 8|ab|}}{2}}. \tag{3.14}$$

From this, it follows that $\rho(\mathcal{H}) < 1$, or $(\mathcal{H}, X^T)$ is intrinsically stable if and only if $|1 - \epsilon| + 2|ab| < 1$. By corollary 3.1, the time-delayed network $(\mathcal{H}, X^T)$ is then stable if this condition holds.

We now turn our attention to determining how modifications of a network's time delays affect the network's stability.

Our first result in this direction states that if a time-delayed dynamical network is known to be intrinsically stable, then removing the network's time delays will not affect the network's stability. This result is novel in the sense that a time-delayed network (or general delayed dynamical system) may become unstable if its delays are removed (see Example 3.2). Hence, an intrinsically stable network has a stronger form of stability, since it remains stable even when its delays are removed.

Following this, we show that it is possible to destabilize a dynamical network $(\mathcal{F}, X)$ by adding delays to the system, even if $\rho(\mathcal{F}) < 1$. Since the delays that destabilize an intrinsically stable network have a specific form, this will enable us to split network delays into two categories. The first class, called a multiple-type delay, comprises those that can have a destabilizing effect on the network. The second class of delays, called single-type delays, consists of those that do not have such an effect.

Before continuing, we formalize the notion of modifying a system's time delays. Our first objective is to describe the removal of delays from a time-delayed dynamical network.

Definition 3.5. Let $(\mathcal{H}, X^T)$ be a time-delayed dynamical network. Then the map $\mathcal{U} : X \to X$ given by

$$\mathcal{U}(\mathbf{x}) = \mathcal{H}(\mathbf{x}, \ldots, \mathbf{x}) \text{ for } \mathbf{x} \in X$$

generates the *undelayed dynamical network* $(\mathcal{U}, X)$, with local systems (Id, X) and interaction $\mathcal{U} : X \to X$.

Remark 3.4. If the local systems (φ, X) are invertible, we can define $(\mathcal{U}, X)$ to be the dynamical network with local systems (φ, X) and interaction

$$U(\mathbf{x}) = H(\mathbf{x}, \varphi^{-1}(\mathbf{x}), \varphi^{-1}(\mathbf{x}), \ldots, \varphi^{-1}(\mathbf{x}))$$

for $\mathbf{x} \in X$. For the sake of generality, though, we define $(\mathcal{U}, X)$ to have the interaction $\mathcal{U} : X \to X$ and no local dynamics, similar to our definition of $(\bar{\mathcal{H}}, X^T)$.

The dynamical network $(\mathscr{U}, X)$ is called the *undelayed version* of $(\mathscr{H}, X^T)$. The following result relates the stability of the undelayed network $(\mathscr{U}, X)$ to the stability of $(\mathscr{H}, X^T)$.

Theorem 3.3. *If $\rho(\mathscr{H}) < 1$, then the undelayed version of this network $(\mathscr{U}, X)$ is intrinsically stable.*

An important consequence of Theorem 3.3 is that if a time-delayed network is intrinsically stable, then the network's stability is invariant under the removal of its time delays. We save the proof of Theorem 3.3 for Sect. 3.3 but illustrate its use with the following example.

Example 3.2. Consider the time-delayed dynamical network $(\mathscr{H}, X^T)$ given by

$$\mathscr{H}(x^k, x^{k-1}) = (\alpha + \gamma)x^k + \alpha x^{k-1},$$

with local system $\varphi(x) = (\alpha + \gamma)x$ and interaction $H(x^k, x^{k-1}) = x^k + \alpha x^{k-1}$, where $X^T = \mathbb{R}^2$ and $\alpha, \gamma \in \mathbb{R}$.

This system has the associated dynamical network

$$\bar{\mathscr{H}}(\mathbf{x}; \alpha, \gamma) = \begin{bmatrix} \bar{\mathscr{H}}_1^0(x_1^0, x_1^{-1}) \\ \bar{\mathscr{H}}_1^{-1}(x_1^0) \end{bmatrix} = \begin{bmatrix} (\alpha + \gamma)x_1^0 + \alpha x_1^{-1} \\ x_1^0 \end{bmatrix},$$

which is linear and therefore stable if the matrix of first partial derivatives

$$D\bar{\mathscr{H}}(\mathbf{0}) = \begin{bmatrix} \alpha + \gamma & \alpha \\ 1 & 0 \end{bmatrix}$$

has a spectral radius strictly within the unit circle.

In contrast, the stability matrix of $(\mathscr{H}, X^T)$ is given by

$$A_{\bar{\mathscr{H}}} = \begin{bmatrix} |\alpha + \gamma| & |\alpha| \\ 1 & 0 \end{bmatrix}.$$

Let $\Omega_1 = \{(\alpha, \gamma) \in \mathbb{R}^2 : \rho(D\bar{\mathscr{H}}(\mathbf{0})) < 1\}$ and $\Omega_2 = \{(\alpha, \gamma) \in \mathbb{R}^2 : \rho(A_{\mathscr{H}}) < 1\}$. Then we have the strict inclusion $\Omega_2 \subset \Omega_1$, which is shown in Fig. 3.1 (left). Consequently, there are parameter values $(\alpha, \gamma) \in \mathbb{R}^2$ for which $(\mathscr{H}, X^T)$ is stable but $\rho(\mathscr{H}) \geq 1$.

Recall that Theorem 3.1 together with lemma 3.1 gives a sufficient condition stating that if $\rho(A_{\mathscr{H}}) < 1$, then $(\mathscr{H}, X^T)$ is stable. However, Theorem 3.3 additionally guarantees that if this condition holds, the undelayed version of this network is also stable. In this example, the undelayed system $(\mathscr{U}, X)$ is given by

$$\mathscr{U}(x; \alpha, \gamma) = (2\alpha + \gamma)x \text{ for } x, \alpha, \gamma \in \mathbb{R},$$

which has a globally attracting fixed point if and only if $|2\alpha + \gamma| < 1$.

3.2 Time-Delayed Dynamical Networks

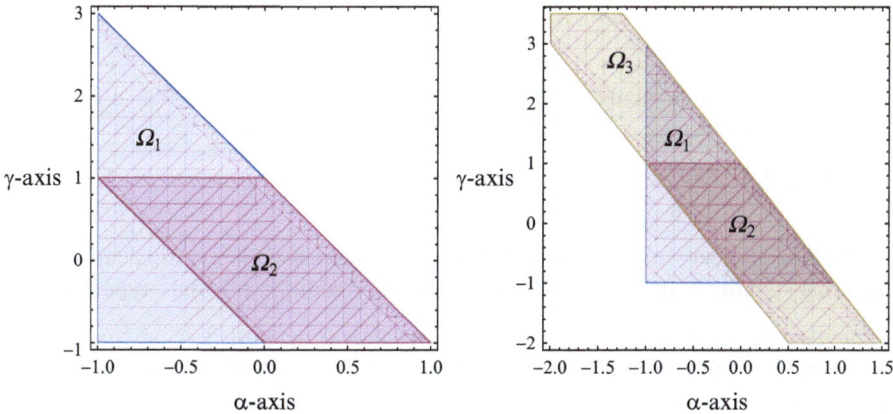

Fig. 3.1 The parameter regions Ω_1 (*blue triangle*), Ω_2 (*red quadrilateral*), and Ω_3 (*yellow strip*) from Example 3.2

With $\Omega_3 = \{(\alpha, \gamma) \in \mathbb{R}^2 : |2\alpha + \gamma| < 1\}$, Fig. 3.1 (right) shows the inclusion $\Omega_2 \subset (\Omega_1 \cap \Omega_3)$. Hence, for every $(\alpha, \gamma) \in \Omega_2$, the time-delayed network $(\mathscr{H}, X^T)$ and its undelayed version $(\mathscr{U}, X)$ are stable, as guaranteed by Theorem 3.3 and Corollary 3.1.

Example 3.2 also points out the following. Since $\Omega_1 \not\subseteq \Omega_3$, it is possible to destabilize a network, or more generally a dynamical system, by removing its time delays. Conversely, given that $\Omega_3 \not\subseteq \Omega_1$, it is also possible to destabilize a system by introducing time delays. Therefore, the converse of Theorem 3.3 does not hold in general. However, this theorem does have a partial converse if we restrict ourselves to specific types of time-delayed interactions.

To demonstrate this, recall from Definition 3.3 that a time-delayed interaction is a function $H : X^T \to X$ given by

$$H_j : \bigoplus_{(i,\tau) \in \mathscr{I}^j} X_i^\tau \to X_j,$$

where $\mathscr{I}^j \subseteq \mathscr{I} \times \mathscr{T}$.

Definition 3.6. The time-delayed network $(\mathscr{H}, X^T)$ is said to have *single-type delays* if for all $j \in \mathscr{I}$, the set $\mathscr{I}^j \subseteq \mathscr{I} \times \mathscr{T}$ has the property

$$(i, \mu) \in \mathscr{I}^j \Rightarrow (i, \nu) \notin \mathscr{I}^j \text{ for all } \nu \neq \mu.$$

If some $(\mathscr{H}, X^T)$ does not have this property, we say that it has a *multiple-type delay*.

Theorem 3.4. *Suppose $(\mathcal{H}, X^T)$ has single-type delays. Then $\rho(\mathcal{H}) < 1$ if and only if $\rho(\mathcal{U}) < 1$.*

As with the other theorems in this section, we prove Theorem 3.4 in Sect. 3.3. Combining Theorems 3.1 and 3.4, we have the following corollary.

Corollary 3.2. *Suppose $(\mathcal{H}, X^T)$ has single-type delays. If either $\rho(\mathcal{H}) < 1$ or $\rho(\mathcal{U}) < 1$, then both $(\mathcal{H}, X^T)$ and $(\mathcal{U}, X)$ are stable.*

Phrased another way, Corollary 3.2 states that if a network is intrinsically stable, then that stability is invariant under the removal or addition of single-type delays.

Example 3.3. Consider the time-delayed dynamical network $(\mathcal{H}, X^T)$ given by equation (3.12) in Example 3.1. The undelayed version of $(\mathcal{H}, X^T)$ is the network $(\mathcal{U}, X)$ given by

$$\mathcal{U}(\mathbf{x}) = \begin{bmatrix} (1-\epsilon)x_1 + 2a \tanh(bx_2) + c_1 \\ (1-\epsilon)x_2 + 2a \tanh(bx_1) + c_2 \end{bmatrix}$$

for $a, b, c_i \in \mathbb{R}$, $X = \mathbb{R}^2$, and $\epsilon \neq 1$.

In computing the undelayed network's stability matrix, we find that

$$A_\mathcal{U} = \begin{bmatrix} |1-\epsilon| & 2|ab| \\ 2|ab| & |1-\epsilon| \end{bmatrix},$$

from which it follows that

$$\rho(\mathcal{U}) = |1-\epsilon| + 2|ab|. \tag{3.15}$$

As can be seen from equation (3.13), the network $(\mathcal{H}, X^T)$ has single-type time delays. Corollary 3.2 then implies that $(\mathcal{H}, X^T)$ is stable if $|1-\epsilon| + 2|ab| < 1$.

Observe that the same condition for the stability of $(\mathcal{H}, X^T)$ was formulated in Example 3.1. However, this equivalence of conditions does not imply that the number of computations required to compute $\rho(\mathcal{H})$ is the same as the number needed to compute $\rho(\mathcal{U})$, as can be seen by comparing equations (3.14) and (3.15). In fact, from a computational point of view, it is always easier to analyze the stability of an undelayed dynamical network $(\mathcal{U}, X)$ than that of the associated time-delayed network $(\mathcal{H}, X^T)$.

If a time-delayed dynamical network $(\mathcal{H}, X^T)$ has a multiple-type delay, then the conclusions of Theorem 3.4 do not necessarily hold. For example, the dynamical network $(\mathcal{U}, X)$ in Example 3.2 has spectral radius $\rho(\mathcal{U}) = 0 < 1$ for $(\alpha, \gamma) = (1, -2)$, although the associated time-delayed network does not have a globally attracting fixed point at these parameter values. The reason is that the time-delayed network has a multiple-type delay.

Combining Theorems 3.3 and 3.4, we see that a network that is intrinsically stable remains stable under the removal of time delays and the introduction of single-type delays. To apply these results to the class of Cohen–Grossberg networks introduced in Sect. 3.1, suppose $(\mathscr{C}, X^T)$ is the time-delayed dynamical network given by

$$\mathscr{C}_j(\mathbf{x}^k, \ldots, \mathbf{x}^{k-(T-1)}) = (1-\epsilon)x_j^{k-\tau_{jj}} + \sum_{i=1}^n W_{ij}\phi(x_i^{k-\tau_{ij}}) + c_j \qquad (3.16)$$

with local systems $\varphi_i(x_i) = (1-\epsilon)x_i + c_i$ and time-delayed interaction

$$C(\mathbf{x}^k, \ldots, \mathbf{x}^{k-(T-1)}) = x_j^{k-\tau_{jj}} + \sum_{i=1}^n W_{ij}\phi\left(\frac{x_i^{k-\tau_{ij}} - c_i}{1-\epsilon}\right). \qquad (3.17)$$

Here, $1 \le i, j \le n$, $0 \le \tau_{ij} \le T-1$, and $\epsilon \ne 1$. As before, $W \in \mathbb{R}^{n \times n}$, $X = \mathbb{R}^n$, and $\phi : \mathbb{R} \to \mathbb{R}$ is a smooth function with Lipschitz constant $\mathscr{L}$.

Equations (3.16) and (3.17) describe a class of time-delayed Cohen–Grossberg networks, whose stability analysis has attracted a good deal of interest [14, 25, 29]. Since $(\mathscr{C}, \mathbb{R}^{nT})$ has single time delays, the following result is a consequence of the proofs of Theorems 3.2 and 3.4.

Theorem 3.5 (Stability of Time-Delayed Cohen–Grossberg Networks). *Suppose $(\mathscr{C}, \mathbb{R}^{nT})$ is the time-delayed Cohen–Grossberg network given by (3.16), where ϕ_i has Lipschitz constant $\mathscr{L}$. If*

$$|1-\epsilon| + \mathscr{L}\rho(|W|) < 1,$$

then $(\mathscr{C}, \mathbb{R}^{nT})$ is intrinsically stable.

Before proving the results in this section, we note that the notion of intrinsic stability of a dynamical network (or more generally, of a dynamical system) is potentially important in the construction of real networks, since such networks are inherently time-delayed. Specifically, if a network is designed to be intrinsically stable, then its stability is much more robust to changes affecting its delays than if it were designed to be only stable.

3.3 Graph Structure of a Dynamical Network

To understand how time delays effect the stability of a dynamical network, we first consider how modifying a network's delays affects the network's graph structure. This will, in turn, leads us to the notion of an isoradial graph (matrix) transformation, which is a transformation that preserves the spectral radius of a graph (matrix). This concept will then allow us to prove Theorems 3.3 and 3.4, found in Sect. 3.2.

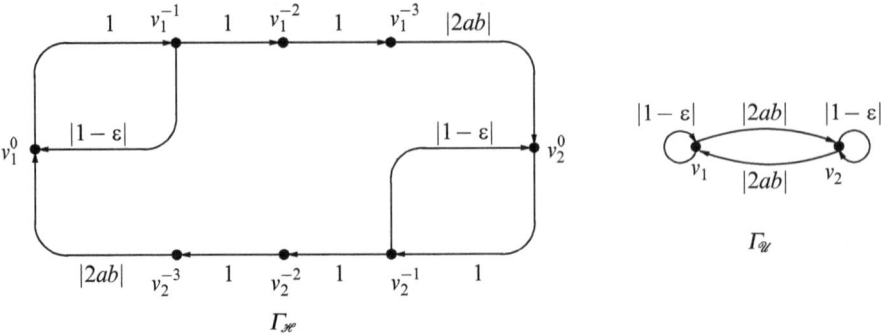

Fig. 3.2 The graph of interactions $\Gamma_{\mathcal{H}}$ and $\Gamma_{\mathcal{U}}$ of the time-delayed network and its undelayed version from Examples 3.1 and 3.3, respectively

To each dynamical network there is an associated weighted directed graph called the network's *graph of interactions*.

Definition 3.7. The graph $\Gamma_{\mathcal{F}} = (V, E, \omega)$ with vertex set $V = \{v_1, \ldots, v_n\}$, edges $E = \{e_{ij} : i \in \mathcal{I}_j, j \in \mathcal{I}\}$, and edge weights $\omega(e_{ij}) = (A_{\mathcal{F}})_{ij}$ is the *graph of interactions* of the dynamical network $(\mathcal{F}, X)$.

The vertex $v_i \in V$ of $\Gamma_{\mathcal{F}}$ corresponds to the ith element of the dynamical network $(\mathcal{F}, X)$. Moreover, there is an edge $e_{ij} \in E$ if and only if the jth component of the interaction $F(\mathbf{x})$ depends on the ith coordinate of $\mathbf{x}$. The weight $\omega(e_{ij})$ measures the strength of the interaction between the ith and jth network elements.

As $(\mathcal{H}, X^T)$ is a dynamical network, it also has a graph of interactions $\Gamma_{\tilde{\mathcal{H}}}$. For convenience, we let this be the graph of interactions of the time-delayed dynamical network $(\mathcal{H}, X^T)$, so that $\Gamma_{\mathcal{H}} = \Gamma_{\tilde{\mathcal{H}}}$.

In this graph, the vertex v_i^{τ} of $\Gamma_{\mathcal{H}}$ corresponds to the component $\tilde{\mathcal{H}}_i^{\tau}$ of $\tilde{\mathcal{H}}$ and represents the ith network element but τ time steps in the past. Each edge from the vertex $v_i^{\tau+1}$ to the vertex v_i^{τ} has unit weight, since $\tilde{\mathcal{H}}_i^{\tau}(x_i^{\tau+1}) = x_i^{\tau+1}$. This corresponds to the fact that these edges simply pass along information until this information is used to update the state of some network element. The graph of interactions of the time-delayed network $(\mathcal{H}, X^T)$ from Example 3.1 and its undelayed version $(\mathcal{U}, X)$ considered in Example 3.3 are shown in Fig. 3.2.

Single and multiple network delays of $(\mathcal{H}, X^T)$ can be characterized in terms of the path and cycle structure of $\Gamma_{\mathcal{H}}$. A time-delayed network $(\mathcal{H}, X^T)$ has single-type delays if there is at most one path of the form $v_i^0, v_i^{-1}, \ldots, v_i^{-\tau+1}, v_j^0$ from v_i^0 to v_j^0 in $\Gamma_{\mathcal{H}}$ for all $i, j \in \mathcal{I}$. If there are two or more such paths, then the network has a multiple-type delay. Using this criterion, the graph structure of $\Gamma_{\mathcal{H}}$ in Fig. 3.2 (left) indicates that the network $(\mathcal{H}, X^T)$ has single-type delays.

We are now in a position to introduce a type of transformation that allows us to modify the structure and spectral radius of a graph in a specific way.

3.3 Graph Structure of a Dynamical Network

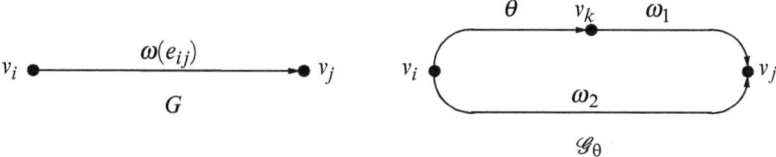

Fig. 3.3 Replacing the edge e_{ij} of G shown left by the subgraph shown right yields the graph $\mathcal{G}_\theta$

Before proceeding, we note that a graph G is a *nonnegative graph* if its adjacency matrix $M(G) \in \mathbb{R}^{n \times n}$ is *nonnegative*, i.e., $M(G)_{ij} \geq 0$ for all $1 \leq i, j \leq n$.

Definition 3.8. For the nonnegative graph $G = (V, E, \omega)$, let $\mathcal{G}_\theta$ be the graph G in which $e_{ij} \in E$ is replaced as in Fig. 3.3, where $\omega_1, \omega_2 \geq 0$, $\omega_1 + \omega_2 = \omega(e_{ij})$, and $\theta > 0$. We call $\mathcal{G}_\theta$ a *bounded radial transformation* of G.

Recall that for every graph G, there is a corresponding matrix $M(G)$, and every matrix is the adjacency matrix of some graph. Therefore, the graph transformation described in Definition 3.8 can equivalently be viewed as a matrix transformation. Before using this transformation, we first consider the following special case.

Lemma 3.2. *Let $\mathcal{G}_\theta$ be a bounded radial transformation of G. If $\theta = \rho(G)$, then $\rho(\mathcal{G}_\theta) = \rho(G)$.*

Lemma 3.2 states that $\mathcal{G}_\theta$ is an *isoradial transformation* of G if $\theta = \rho(G)$, i.e., a transformation that preserves the spectral radius of G. In order to prove Lemma 3.2, we use the Perron–Frobenius theorem and the following standard terminology.

A directed graph G is called *strongly connected* if there is a path from each vertex to every other vertex in G or if G consists of a single vertex. The *strongly connected components* of G are its maximal strongly connected subgraphs. If $M \in \mathbb{R}^{n \times n}$, then M is said to be *irreducible* if the graph G with adjacency matrix M is strongly connected.

Theorem 3.6. *(Perron-Frobenius) Let $M \in \mathbb{R}^{n \times n}$ and suppose that M is irreducible and nonnegative. Then*

(a) $\rho(M) > 0$;
(b) $\rho(M)$ is an eigenvalue of M;
(c) $\rho(M)$ is an algebraically simple eigenvalue of M; and
(d) the left and right eigenvectors x and y associated with $\rho(M)$ have strictly positive entries.

If a graph G is not strongly connected, then it has strongly connected components $\mathbb{S}(G)_1 \ldots, \mathbb{S}(G)_N$. For $M = M(G)$, let M_j be the adjacency matrix of the graph $\mathbb{S}_j(G)$. Then the eigenvalues of M are

$$\sigma(M) = \bigcup_{j=1}^{N} \sigma(M_j), \tag{3.18}$$

from which it follows that

$$\rho(M) = \max_{1 \le j \le N} \rho(M_j). \tag{3.19}$$

We call a strongly connected component $\mathbb{S}_j(G)$ *trivial* if it consists of a single vertex without a loop, in which case $\sigma(\mathbb{S}_j(G)) = \{0\}$. We now give a proof of Lemma 3.2.

Proof. Let $\mathscr{G} = \mathscr{G}_\theta$ be a bounded radial transformation of $G = (V, E, \omega)$ in which $\theta = \rho(G)$, $V = \{v_2, \ldots, v_n\}$, and the edge $e_{23} \in E$ is replaced as in Fig. 3.3, where $v_k = v_1$. Let $M(\mathscr{G}) - \theta I$ have the block form

$$M(\mathscr{G}) - \theta I = \begin{bmatrix} A & B \\ C & D \end{bmatrix},$$

where A is a 1×1 matrix. Assuming that G is strongly connected, then $A = [\theta]$ is invertible by part (a) of the Perron–Frobenius theorem. Using the identity

$$\det \begin{bmatrix} A & B \\ C & D \end{bmatrix} = \det(A) \cdot \det(D - CA^{-1}B),$$

we have $\det(M(\mathscr{G}) - \theta I) = \theta \det(D - \theta^{-1}CB)$, where

$$(D - \theta^{-1}CB)_{ij} = \begin{cases} \omega_1 + \omega_2 & \text{for } i = 2, j = 3, \\ D_{ij} & \text{otherwise.} \end{cases}$$

Since $\omega_1 + \omega_2 = \omega(e_{23})$, it follows that $D - \theta^{-1}CB = M(G) - \theta I$, so that

$$\det(M(\mathscr{G}) - \theta I) = \theta \det(M(G) - \theta I) = 0,$$

which implies $\theta \in \sigma(\mathscr{G})$.

With $\mathscr{G}$ written as a function of ω_1, the claim is that $\rho(\mathscr{G}(\omega_1)) = \theta$ for $\omega_1 \in [0, \omega(e_{ij})]$. To verify this, we note that $\mathscr{G}(0)$ has the strongly connected components G and v_1, where v_1 is trivial. Hence, equation (3.19) implies $\rho(\mathscr{G}(0)) = \theta$. Since $\theta \in \sigma(\mathscr{G}(\omega_1))$, and $\rho(\mathscr{G}(\omega_1))$ is continuous with respect to ω_1 for $\omega_1 \in [0, \omega(e_{ij})]$, parts (b) and (c) of the Perron–Frobenius theorem imply $\rho(\mathscr{G}(\omega_1)) = \theta$. This verifies the claim and completes the proof. □

The reason the graph $\mathscr{G}_\theta$ is called a *bounded radial transform* in Definition 3.8 is due to the following result.

Lemma 3.3. *If G is nonnegative, then $\rho(G) < \rho(\mathscr{G}_\theta) < \theta$ if and only if $\theta > \rho(G)$.*

That is, the spectral radius of the graph $\mathscr{G}_\theta$ is bounded by θ as long as $\theta > \rho(G)$. To prove Lemma 3.3, we require the following result, which can be found, for instance, in [22].

3.3 Graph Structure of a Dynamical Network

Proposition 3.3. *Let $A, B \in \mathbb{R}^{n \times n}$ be nonnegative matrices and suppose that A is irreducible. Then $\rho(A + B) > \rho(A)$ if $B \neq 0$.*

That is, the spectral radius of a nonnegative irreducible matrix is strictly monotone in each of its entries. This allows us to give the following proof of Lemma 3.3.

Proof. Let $\mathcal{M}_\theta = M(\mathcal{G}_\theta)$ and $M = M(G)$. Supposing G is nonnegative and strongly connected, then $\mathcal{M}_{\rho(M)}$ is nonnegative and irreducible. Moreover, $\mathcal{M}_{\rho(M)}$ is the adjacency matrix of the isoradial transformation $\mathcal{G}$, so Lemma 3.2 implies that $\rho(\mathcal{M}_{\rho(M)}) = \rho(M)$. This together with Proposition 3.3 implies that for $\gamma > 1$,

$$\rho(\mathcal{M}_{\rho(M)}) < \rho(\mathcal{M}_{\gamma\rho(M)}) < \rho(\gamma \mathcal{M}_{\rho(M)}).$$

Since $\rho(\gamma \mathcal{M}_{\rho(M)}) = \gamma \rho(M)$, then letting $\theta = \gamma \rho(M)$, we have

$$\rho(M) < \rho(\mathcal{M}_\theta) < \theta \text{ if and only if } \theta > \rho(M)$$

since $\theta > 0$. This implies the result when G is strongly connected.

Suppose, then, that G is nonnegative but is not strongly connected. In this case, G can be decomposed into its strongly connected components $\mathbb{S}(G)_1, \ldots, \mathbb{S}(G)_N$. By way of notation, we let M_m denote the adjacency matrix of the strongly connected component $\mathbb{S}(G)_m$ for $1 \leq m \leq N$. Note that since G is nonnegative, each M_m is nonnegative.

If e_{ij} is an edge of $\mathbb{S}(G)_\ell$, then we consider two cases. First, suppose $\mathbb{S}(G)_\ell$ is nontrivial. Then the graph $\mathcal{G}_\theta$ with adjacency matrix $\mathcal{M}_\theta$ has the strongly connected components

$$\mathbb{S}(G)_1 \ldots, \mathbb{S}_\theta(G)_\ell, \ldots, \mathbb{S}(G)_N,$$

where we let $(\mathcal{M}_\theta)_\ell$ denote the nonnegative adjacency matrix of $\mathbb{S}_\theta(G)_\ell$. Since the adjacency matrix of a nontrivial strongly connected component is irreducible, it follows that $\rho((\mathcal{M}_\theta)_\ell) < \theta$ if and only if $\rho(M) < \theta$.

From equation (3.19), we have

$$\rho(\mathcal{M}_\theta) = \max\{\rho(M_1), \ldots, \rho((\mathcal{M}_\theta)_\ell), \ldots, \rho(M_N)\} = \max\{\rho(M), \rho((\mathcal{M}_\theta)_\ell)\}.$$

Hence, if $\rho(M) < \theta$, then $\rho((\mathcal{M}_\theta)_\ell) < \theta$, implying $\rho(\mathcal{M}_\theta) < \theta$. Conversely, if $\rho(\mathcal{M}_\theta) < \theta$, then $\rho(M) < \theta$, and the result holds in this case.

If $\mathbb{S}(G)_\ell$ is trivial, then $\omega_1, \omega_2 = 0$, and $\mathcal{M}_\theta$ has the strongly connected components

$$\mathbb{S}(G)_1, \ldots, \mathbb{S}(G)_N, \mathbb{S}(G)_{N+1},$$

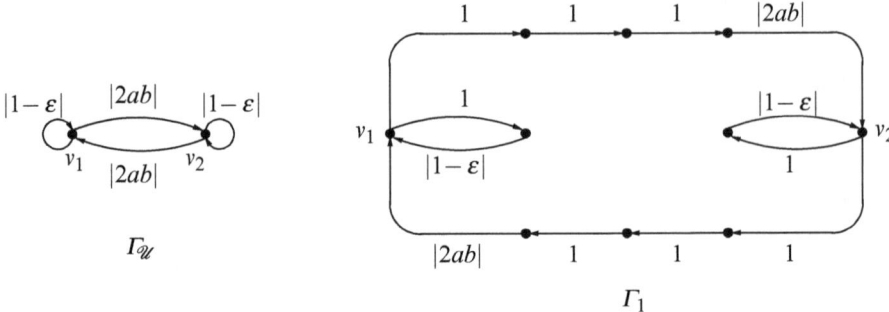

Fig. 3.4 A transform Γ_1 of the graph $\Gamma_\mathscr{U}$ from Example 3.3, via a sequence of bounded radial transformations

where $\mathbb{S}(G)_{N+1}$ is trivial with single vertex v_k. Hence, $\sigma(M_{N+1}) = \{0\}$, implying $\rho(\mathscr{M}_\theta) = \rho(M)$ by equation (3.18). Hence the result follows for the case in which the edge e_{ij} is an edge of the component $\mathbb{S}(G)_\ell$.

If $v_i \in \mathbb{S}(G)_\ell$ and $v_j \in \mathbb{S}(G)_m$ for $\ell \neq m$, then the graphs associated with M and $\mathscr{M}_\theta$ have the same nontrivial strongly connected components. Equation (3.19) then implies that $\rho(\mathscr{M}_\theta) = \rho(M)$, and the result follows for the case in which e_{ij} does not belong to a strongly connected component. Since this exhausts all cases, this completes the proof. □

Example 3.4. Consider the graph of interactions $\Gamma_\mathscr{U}$ shown in Fig. 3.4 (left) of the undelayed network $(\mathscr{U}, X)$ from Example 3.3. By a series of bounded radial transformations, it is possible to transform $\Gamma_\mathscr{U}$ into the graph Γ_1 shown on the right. This is done at each step by letting $\theta = 1$ and $\omega_2 = 0$, so that the edge being modified is bisected, with one edge retaining the original edge weight and the other receiving the weight 1. Importantly, we note that Lemma 3.3 implies that $\rho(\mathscr{U}) < 1$ if and only if $\rho(\Gamma_1) < 1$.

Although the graph Γ_1 in Fig. 3.4 is not the graph of interactions $\Gamma_\mathscr{H}$ of the network $(\mathscr{H}, X^T)$ shown in fig. 3.2 (left), the two have a similar structure. To quantify in what sense two graphs are similar, we introduce the following.

Definition 3.9. For the graph $G = (V, E, \omega)$, suppose $S \in st_0(G)$ such that each vertex of V belongs to a branch of $\mathscr{B}_S(G)$. Then we call S a *complete branching set* of G with respect to S.

For a graph $G \in \mathbb{G}$, we let $br_0(G)$ denote the collection of complete branching sets of the graph G. The single difference between a complete structural set S and a complete branching set T is that the branching set $\mathscr{B}_S(G)$ may not contain every vertex of G, whereas $\mathscr{B}_T(G)$ does.

This can be seen, for instance, in Fig. 3.5. Here, the graph G has the complete structural set $S = \{v_2, v_3\}$. The set S, however, is not a complete branching set,

3.3 Graph Structure of a Dynamical Network

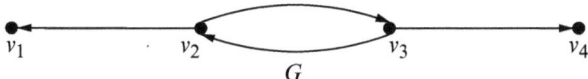

Fig. 3.5 The graph G has the complete structural set $S = \{v_2, v_3\}$ and the complete branching set $T = \{v_1, v_2, v_3, v_4\}$

since no branch of $\mathscr{B}_S(G) = \{v_2, v_3; v_3, v_2\}$ contains either v_1 or v_4. In fact, the only complete branching set of G is the set $T = \{v_1, v_2, v_3, v_4\}$.

Definition 3.10. For the graph $G = (V, E, \omega)$, suppose $S \in br_0(G)$. Then the graph $\mathscr{X}_S(G)$ is called an *isoradial expansion* of G with respect to S.

Recall from Sect. 2.5 that for some $S \in st(G)$, an isospectral graph expansion $\mathscr{X}_S(G)$ is the graph G in which the branches of $\mathscr{B}_S(G)$ have been made independent. An isoradial expansion of a graph G is then an isospectral expansion with respect to a special type of structural set S, namely, one that is both a complete structural set and has the property that each vertex of G belongs to some branch of $\mathscr{B}_S(G)$.

As a corollary to Corollary 2.5, we have the following result.

Corollary 3.3. *Let $G = (V, E, \omega)$ and $S \in br_0(G)$. Then $\rho(\mathscr{X}_S(G)) = \rho(G)$.*

Proof. Suppose $G = (V, E, \omega)$ and $S \in br_0(G)$. Since each cycle of G including loops must contain a vertex in S, it follows that $\omega(e_{ii}) = 0$ for each $i \notin \mathscr{I}_S$.

Via Theorem 2.8, it then follows that

$$\det\left(M(\mathscr{X}_S(G)) - \lambda I\right) = \det\left(M(G) - \lambda I\right) \prod_{v_i \in V-S} (-\lambda)^{n_i - 1},$$

where n_i is the number of branches in $\mathscr{B}_S(G)$ containing $v_i \in V$. Hence $\sigma(G)$ and $\sigma(\mathscr{X}_S(G))$ differ by at most some number of zeros, implying that the two graphs have the same spectral radius. □

Example 3.5. Consider the graph $\Gamma_{\mathscr{H}}$ shown in Fig. 3.6 (left), which is the graph of interactions of the time-delayed network $(\mathscr{H}, X^T)$ from Example 3.1. Since each cycle of $\Gamma_{\mathscr{H}}$ passes through a vertex of $S = \{v_1^0, v_2^0\}$ and each vertex of this graph belongs to a cycle, we have $S \in br_0(\Gamma_{\mathscr{H}})$. The isoradial expansion $\mathscr{X}_S(\Gamma_{\mathscr{H}})$ is shown in Fig. 3.6 (right), which has spectral radius $\rho(\mathscr{X}_S(\Gamma_{\mathscr{H}})) = \rho(\Gamma_{\mathscr{H}})$ by Corollary 3.3.

Note that the graph $\mathscr{X}_S(\Gamma_{\mathscr{H}})$ is identical to the graph Γ_1 from Example 3.4, which was shown to have spectral radius $\rho(\Gamma_1) < 1$ if and only if $\rho(\mathscr{U}) < 1$. Hence $\rho(\mathscr{U}) < 1$ if and only if $\rho(\mathscr{H}) < 1$, where $(\mathscr{U}, X)$ is the undelayed version of the time-delayed network $(\mathscr{H}, X^T)$. That is, using isoradial expansions together with bounded radial transformations, it is possible to show that the stability of a delayed network implies the stability of the network without delays, and vice versa.

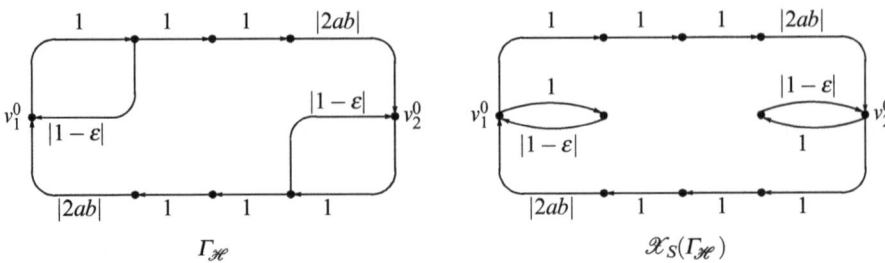

Fig. 3.6 The graph $\Gamma_{\mathcal{H}}$ from Fig. 3.2 and its branch expansion $\mathcal{X}_S(\Gamma_{\mathcal{H}})$ over the complete branching set $S = \{v_1^0, v_2^0\}$

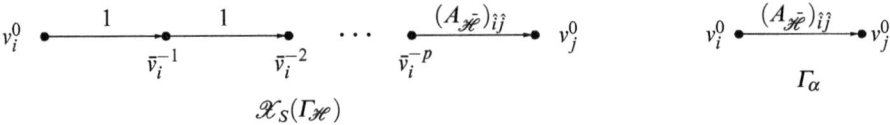

Fig. 3.7 The branch α in $\mathcal{X}_S(\Gamma_{\mathcal{H}})$ (left) and its replacement by a single edge in Γ_α (right)

Our goal in the remainder of this section is to use isoradial expansions and bounded radial transformations to prove Theorems 3.3 and 3.4.

A proof of Theorem 3.4 is the following.

Proof. Let $(\mathcal{H}, X^T)$ be a time-delayed network. The claim is that $S = \{v_i^0 : i \in \mathcal{I}\}$ is a complete branching set of $\Gamma_{\mathcal{H}}$. To see this, note that if $v_1, \ldots, v_m$ is a path or cycle in $\Gamma_{\mathcal{H}}$ containing no vertices of S, then (3.10) implies that there is some $i \in \mathcal{I}$ such that $v_\ell = v_i^{\tau-\ell}$ for $1 \leq \ell \leq m$. Since each v_ℓ is distinct, then in order to form a cycle of $\Gamma_{\mathcal{H}}$, we need an element of S. Moreover, each vertex of $\Gamma_{\mathcal{H}}$ belongs to a path (cycle) of the form $v_i^0, v_i^{-1}, \ldots, v_i^{-\tau}, v_j^0$ (where $i = j$). Hence, $S \in br_0(\Gamma_{\mathcal{H}})$.

Suppose $(\mathcal{H}, X^T)$ has single-type delays. Then for $i, j \in \mathcal{I}$, there is at most one branch in $\mathcal{B}_S(\Gamma_{\mathcal{H}})$ from v_i^0 to v_j^0, which implies that the same is true of the isoradial expansion $\mathcal{X}_S(\Gamma_{\mathcal{H}})$. If $\alpha = v_i^0, \bar{v}_i^{-1}, \ldots, \bar{v}_i^{-p}, v_j^0$ is a branch in $\mathcal{B}_S(\mathcal{X}_S(\Gamma_{\mathcal{H}}))$, then it must have the form shown in Fig. 3.7 (left), where $\hat{i} = (i, -p)$ and $\hat{j} = (j, 0)$ are in $\mathcal{I} \times \mathcal{T}$.

Via a sequence of bounded radial transformations with $\omega_1 = (A_{\tilde{\mathcal{H}}})_{\hat{i}\hat{j}}$, $\omega_2 = 0$, and $\theta = 1$, the single edge of Γ_α shown in Fig. 3.7 (right) can be replaced by the branch α. Lemma 3.3 then implies that $\rho(\Gamma_\alpha) < 1$ if and only if $\rho(\mathcal{X}_S(\Gamma_{\mathcal{H}})) < 1$. For the same reason, replacing each $\beta = v_\ell^0, v_\ell^{-1}, \ldots, v_\ell^{-q}, v_m^0 \in \mathcal{B}_S(\mathcal{X}_S(\Gamma_{\mathcal{H}}))$ by a single edge from v_ℓ^0 to v_m^0 with weight $(A_{\tilde{\mathcal{H}}})_{\hat{m}\hat{q}}$ will result in the graph Γ with the property

$$\rho(\Gamma) < 1 \text{ if and only if } \rho(\mathcal{X}_S(\Gamma_{\mathcal{H}})) < 1. \qquad (3.20)$$

3.3 Graph Structure of a Dynamical Network

For $i, j \in \mathscr{I}$ the adjacency matrix $M(\Gamma)$ of Γ has entries

$$M(\Gamma)_{ij} = (A_{\tilde{\mathscr{H}}})_{\hat{i}\hat{j}} = \max_{\mathbf{x} \in X} |(D\tilde{\mathscr{H}})_{\hat{j}\hat{i}}(\mathbf{x}, \dots, \mathbf{x})| = \max_{\mathbf{x} \in X} |(D\mathscr{U})_{ji}(\mathbf{x})| = (A_{\mathscr{U}})_{ij}, \tag{3.21}$$

where the second-to-last equality follows from the fact that (H, X^T) has only single-type delays.

Thus, $\rho(\Gamma) = \rho(\mathscr{U})$. The reason that we have

$$(A_{\tilde{\mathscr{H}}})_{\hat{i}\hat{j}} = \max_{\mathbf{x} \in X} |(D\tilde{\mathscr{H}})_{\hat{j}\hat{i}}(\mathbf{x}, \dots, \mathbf{x})| \text{ and } \max_{\mathbf{x} \in X} |(D\mathscr{U})_{ji}(\mathbf{x})| = (A_{\mathscr{U}})_{ij}$$

is that both $(\tilde{\mathscr{H}}, X^T)$ and $(\mathscr{U}, X)$ are assumed to have no local dynamics. The fact that $\rho(\mathscr{H}) = \rho(\mathscr{X}_S(\Gamma_{\mathscr{H}}))$, via Corollary 3.3, together with (3.20) implies that $\rho(\mathscr{U}) < 1$ if and only if $\rho(\mathscr{H}) < 1$. This completes the proof. □

We now give a proof of Theorem 3.3.

Proof. For the time-delayed network $(\mathscr{H}, X^T)$, it follows from the proof of Theorem 3.4 that $S = \{v_i^0 : i \in \mathscr{I}\}$ is a complete branching set of $\Gamma_{\mathscr{H}}$. Suppose, then, that there is a single branch $\alpha = v_i^0, \bar{v}_i^{-1}, \dots, \bar{v}_i^{-p}, v_j^0$ in $\mathscr{B}_S(\mathscr{X}_S(\Gamma_{\mathscr{H}}))$ from v_i^0 to v_j^0, as shown in Fig. 3.7 (left) with $\hat{i} = (i, -p)$ and $\hat{j} = (0, j)$ in $\mathscr{I} \times \mathscr{T}$.

Following the proof of Theorem 3.4, the branch α can be replaced with a single edge, as in Fig. 3.7 (right), such that

$$\rho(\Gamma_\alpha) < 1 \text{ if and only if } \rho(\mathscr{X}_S(\Gamma_{\mathscr{H}})) < 1.$$

Suppose, then, that there are two branches

$$\alpha = v_i^0, \bar{v}_i^{-1}, \dots, \bar{v}_i^{-p}, v_j^0, \quad \beta = v_i^0, \tilde{v}_i^{-1}, \dots, \tilde{v}_i^{-q}, v_j^0$$

in $\mathscr{B}_S(\mathscr{X}_S(\Gamma_{\mathscr{H}}))$ from v_i^0 to v_j^0, as shown in Fig. 3.8 (left), where $\bar{i} = (i, -p)$ and $\tilde{i} = (i, -q)$ are in $\mathscr{I} \times \mathscr{T}$.

By repeated use of Lemma 3.3 with $\omega_2 = 0$ and $\theta = 1$, the branch α can be replaced with a single edge and β by two edges, as in the intermediate graph Γ_{int} in Fig. 3.8 (center). By another use of Lemma 3.3 with $\omega_1 = (A_{\tilde{\mathscr{H}}})_{\bar{i}\hat{j}}$, $\omega_2 = (A_{\tilde{\mathscr{H}}})_{\tilde{i}\hat{j}}$, and $\theta = 1$, it follows that the graph $\Gamma_{\alpha/\beta}$ shown in Fig. 3.8 (right) has the property

$$\rho(\Gamma_{\alpha/\beta}) < 1 \text{ if and only if } \rho(\mathscr{X}_S(\Gamma_{\mathscr{H}})) < 1.$$

Continuing in this manner, suppose there are branches $\alpha_1, \dots, \alpha_m \in \mathscr{B}_S(\mathscr{X}_S(\Gamma_{\mathscr{H}}))$ from v_i^0 to v_j^0. Then, through a sequence of bounded radial transformations, these branches can be replaced by a single edge with weight

$$\omega_{ij} = \sum_{\ell=1}^m (A_{\tilde{\mathscr{H}}})_{i_\ell, \hat{j}} \text{ where } i_\ell = (i, |\alpha_\ell| + 1) \in \mathscr{I} \times \mathscr{T}.$$

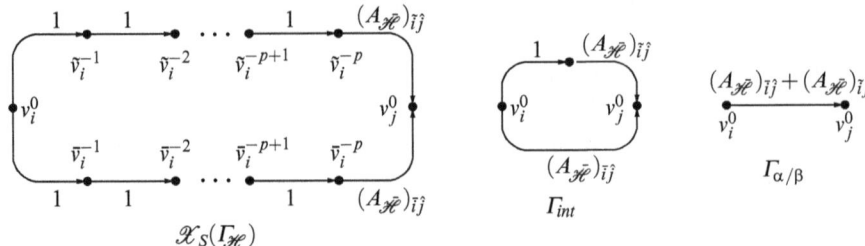

Fig. 3.8 The branches α and β in $\mathscr{X}_S(\Gamma_{\mathscr{H}})$ (left) and their replacements in Γ_{int} (center) and $\Gamma_{\alpha/\beta}$ (right)

The claim is that $(A_{\mathscr{U}})_{ij} < \omega_{ij}$, where $(\mathscr{U}, X)$ is the undelayed version of $(\mathscr{H}, X^T)$. Indeed, since $\alpha_1, \ldots, \alpha_m \in \mathscr{B}_S(\mathscr{X}_S(\Gamma_{\mathscr{H}}))$ are the branches of $\mathscr{X}_S(\Gamma_{\mathscr{H}})$ from v_i^0 to v_j^0, the function $\mathscr{H}_j : X^T \to X$ depends on the variables indexed by i of the form

$$x_i^{k-|\alpha_\ell|+1} \text{ for } 1 \leq \ell \leq m.$$

Therefore, $(A_{\mathscr{U}})_{ij} \leq \omega_{ij}$, since

$$\max_{\mathbf{x} \in X} |(D\mathscr{U})_{ji}(\mathbf{x})| = \max_{\mathbf{x} \in X} |(D\mathscr{H})_{j,i_\ell}(\mathbf{x}, \ldots, \mathbf{x})| \leq \max_{\mathbf{x} \in X^T} \sum_{\ell=1}^m |(D\bar{\mathscr{H}})_{\hat{j},i_\ell}(\mathbf{x})| = \omega_{ij}.$$

(3.22)

Let Γ be the graph in which each set of branches in $\mathscr{X}_S(\Gamma_{\mathscr{H}})$ from v_i^0 to v_j^0 is replaced by the edge with weight ω_{ij} for all $i, j \in \mathscr{I}$. Then equation (3.22) implies that $\rho(\mathscr{U}) \leq \rho(\Gamma)$.

Since Γ can be transformed via a sequence of bounded radial transformations with $\theta = 1$ into $\mathscr{X}_S(\Gamma_{\mathscr{H}})$, Lemma 3.3 implies that $\rho(\Gamma) < 1$ if and only if $\rho(\mathscr{X}_S(\Gamma_{\mathscr{H}})) < 1$. The fact that $\rho(\mathscr{X}_S(\Gamma_{\mathscr{H}})) = \rho(\Gamma_{\mathscr{H}})$ implies that if $\rho(\mathscr{H}) < 1$, then $\rho(\mathscr{U}) < 1$, completing the proof. □

3.4 Implicit Delays and Restrictions of Dynamical Networks

One of the major obstacles to understanding the dynamics of a network (high-dimensional dynamical system) is that the information needed to do so is spread throughout the various network elements and interactions (system components). In this section, we consider whether it is possible to consolidate this information to gain improved estimates of a network's stability.

Suppose we consider two different elements of a dynamical network. Even if there is no direct interaction between the two, one may influence the dynamics of

3.4 Implicit Delays and Restrictions of Dynamical Networks

the other after some number of time steps $\tau > 1$. This amounts to an implicit time-delayed interaction between these elements.

In terms of the graph structure of $(\mathscr{F}, X)$, there is an *implicit time-delayed interaction* from the ith to the jth network element if there is a path in $\Gamma_\mathscr{F}$ from v_i to v_j. By choosing a particular subset of network elements, the claim is that it is possible to associate a time-delayed network with $(\mathscr{F}, X)$ whose delays represent the implicit delays among this collection of network elements.

To construct this time-delayed network, let $\mathbf{x}^{k+1} = \mathscr{F}(\mathbf{x}^k)$, and suppose that the set S is in $br_0(\Gamma_\mathscr{F})$. For $j \in \mathscr{I}_S$, let $\mathscr{F}_{j,0}(\mathbf{x}^k) = \mathscr{F}_j(\mathbf{x}^k)$. For $\ell \geq 1$, we recursively define $\mathscr{F}_{j,\ell} = \mathscr{F}_{j,\ell}(\mathbf{x}^k, \ldots, \mathbf{x}^{k-\ell})$ to be the function

$$\mathscr{F}_{j,\ell-1} = \mathscr{F}_{j,\ell-1}(\mathbf{x}^k, \ldots, \mathbf{x}^{k-\ell+1}),$$

in which the variable $x_i^{k-\ell+1}$ is replaced by the function $\mathscr{F}_i(\mathbf{x}^{k-\ell})$ for all $i \notin \mathscr{I}_S$.

If for some $m \geq 0$, each variable of $\mathscr{F}_{j,m}(\mathbf{x}^k, \ldots, \mathbf{x}^{k-m})$ is indexed by an element of $\mathscr{I}_S$, then we let

$$\mathscr{H}_S \mathscr{F}_j(\mathbf{x}_S^k, \ldots, \mathbf{x}_S^{k-m}) = \mathscr{F}_{j,m}(\mathbf{x}_S^k, \ldots, \mathbf{x}_S^{k-m}), \tag{3.23}$$

where each $\mathbf{x}_S^{k-\tau}$ belongs to $X_S = X|_{\mathscr{I}_S}$.

By choosing S to be a complete branching set, we are guaranteed that the function $\mathscr{H}_S \mathscr{F}_j$ exists, i.e., that there is some $m < \infty$ such that each variable of $\mathscr{F}_{j,m}$ is indexed by an element of $\mathscr{I}_S$. The reason for this is that otherwise, there would be a cycle of $\Gamma_\mathscr{F}$ containing no element of the structural set S.

Definition 3.11. For $S \in br_0(\Gamma_\mathscr{F})$, let $(\mathscr{H}_S \mathscr{F}, X_S^T)$ be the dynamical network with components defined by equation (3.23) for $j \in \mathscr{I}_S$. This network is called the *time-delayed version* of $(\mathscr{F}, X)$ with respect to S.

The time-delayed network $(\mathscr{H}_S \mathscr{F}, X_S^T)$ is considered to be a network with no local dynamics, since the compositions needed to construct this system do not, in general, allow for a decomposition into an interaction and local systems.

Example 3.6. Consider the dynamical network $(\mathscr{F}, X)$ given by

$$\mathscr{F}(\mathbf{x}; c) = \begin{bmatrix} \tanh(x_6) + c \\ \tanh(x_1) + c \\ \tanh(x_2) + \tanh(x_5) + c \\ \tanh(x_3) + c \\ \tanh(x_4) + c \\ \tanh(x_2) + \tanh(x_5) + c \end{bmatrix}, \quad \mathbf{x} \in \mathbb{R}^6$$

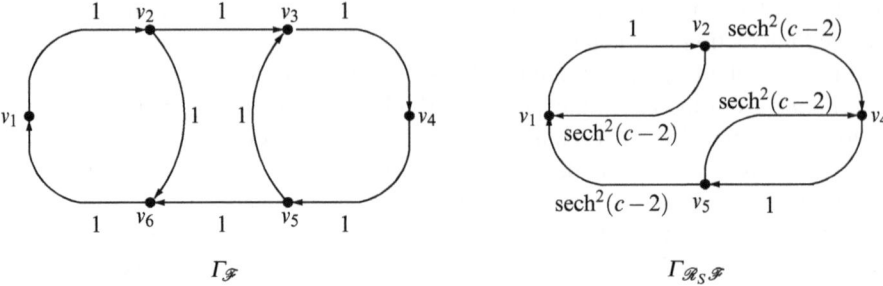

Fig. 3.9 The graph $\Gamma_{\mathscr{F}}$ of the network in Example 3.6 and the graph of its restriction $\Gamma_{\mathscr{R}_S \mathscr{F}}$ to $S = \{v_1, v_2, v_4, v_5\}$

with local systems $\varphi_i(x_i) = \tanh(x_i)$ and interaction

$$F(\mathbf{x}) = \begin{cases} x_{i-1} + c & \text{for } i = 1, 2, 4, 5, \\ x_2 + x_5 + c & \text{for } i = 3, 6, \end{cases}$$

where the indices are taken modulo 6, $X = \mathbb{R}^6$, and $c \in \mathbb{R}$. The network $(\mathscr{F}, X)$ can be thought of as an undelayed Cohen–Grossberg network, where we allow $\epsilon = 1$.

Formally, $(\mathscr{F}, X)$ can be written as the time-delayed network $\mathbf{x}^{k+1} = \mathscr{F}(\mathbf{x}^k)$ with

$$\mathbf{x}^{k+1} = \begin{bmatrix} \tanh(x_6^k) + c \\ \tanh(x_1^k) + c \\ \tanh(x_2^k) + \tanh(x_5^k) + c \\ \tanh(x_3^k) + c \\ \tanh(x_4^k) + c \\ \tanh(x_2^k) + \tanh(x_5^k) + c \end{bmatrix}, \; \mathbf{x}^k \in \mathbb{R}^6.$$

As can be seen in Fig. 3.9, the vertex set $S = \{v_1, v_2, v_4, v_5\}$ is a complete branching set of $\Gamma_{\mathscr{F}}$. To construct $(\mathscr{H}_S \mathscr{F}, X_S^T)$, we let $\mathscr{F}_{j,0}(\mathbf{x}^k) = \mathscr{F}_j(\mathbf{x}^k)$ for each $j \in \mathscr{I}_S$ and note that $\mathscr{H}_S \mathscr{F}_j(\mathbf{x}^k) = \mathscr{F}_{j,0}(\mathbf{x}^k)$ for $j = 2, 5$, since each variable of these two functions is indexed by an element of $\mathscr{I}_S$. Replacing the variables x_3^k and x_6^k by $\mathscr{F}_3(\mathbf{x}^{k-1})$ and $\mathscr{F}_6(\mathbf{x}^{k-1})$, respectively, in $\mathscr{F}_{j,0}(\mathbf{x}^k)$ for $j = 1, 4$ yields the function

$$\mathscr{F}_{j,1}(\mathbf{x}^k, \mathbf{x}^{k-1}) = \tanh(\tanh(x_2^{k-1}) + \tanh(x_5^{k-1}) + c) + c.$$

Since each of this function's variables is indexed by an element of $\mathscr{I}_S$, the function $\mathscr{H}_S \mathscr{F}_j(\mathbf{x}^k, \mathbf{x}^{k-1})$ is equal to $\mathscr{F}_{j,1}(\mathbf{x}^k, \mathbf{x}^{k-1})$ for $j = 1, 4$.

3.4 Implicit Delays and Restrictions of Dynamical Networks

The time-delayed network $(\mathcal{H}_S\mathcal{F}, X_S^T)$ associated with $(\mathcal{F}, X)$ is then given by

$$\mathbf{x}^{k+1} = \begin{bmatrix} \mathcal{H}_S\mathcal{F}_1(x_2^{k-1}, x_5^{k-1}) \\ \mathcal{H}_S\mathcal{F}_2(x_1^k) \\ \mathcal{H}_S\mathcal{F}_4(x_2^{k-1}, x_5^{k-1}) \\ \mathcal{H}_S\mathcal{F}_5(x_4^k) \end{bmatrix} = \begin{bmatrix} \tanh(\tanh(x_2^{k-1}) + \tanh(x_5^{k-1}) + c) + c \\ \tanh(x_1^k) + c \\ \tanh(\tanh(x_2^{k-1}) + \tanh(x_5^{k-1}) + c) + c \\ \tanh(x_4^k) + c \end{bmatrix},$$
(3.24)

where $T = 2$. The system $(\mathcal{H}_S\mathcal{F}, X_S^T)$ is the network $(\mathcal{F}, X)$, in which the implicit delays between the elements indexed by S are made explicit.

When $(\mathcal{F}, X)$ is written in the form of a delayed network $\mathbf{x}^{k+1} = \mathcal{F}(\mathbf{x}^k)$, the initial condition $\mathbf{x}^0 \in X$ has the orbit $\{\mathbf{x}^k\}_{k \geq 0}$ under the action of $\mathcal{F}$. For $\mathbf{x}^0 \in X$, it follows from this construction that

$$\mathcal{F}_j(\mathbf{x}^k) = \mathcal{H}_S\mathcal{F}_j(\mathbf{x}_S^k, \ldots, \mathbf{x}_S^{k-T+1})$$

for each $j \in \mathcal{I}_S$ and $k \geq T - 1$. That is, $(\mathcal{H}_S\mathcal{F}, X_S^T)$ and $(\mathcal{F}, X)$ have the same dynamics when restricted to S.

Since every vertex of $\Gamma_\mathcal{F}$ belongs to a branch of $\mathcal{B}_S(\Gamma_\mathcal{F})$, this implies the following result.

Lemma 3.4. *For $S \in br_0(\Gamma_\mathcal{F})$, the time-delayed network $(\mathcal{H}_S\mathcal{F}, X_S^T)$ is stable if and only if $(\mathcal{F}, X)$ is stable.*

By constructing the time-delayed dynamical network $(\mathcal{H}_S\mathcal{F}, X_S^T)$, we absorb the dynamics of the elements of $(\mathcal{F}, X)$, that are not indexed by $\mathcal{I}_S$, into the time-delayed interaction of $(\mathcal{H}_S\mathcal{F}, X_S^T)$. In this sense, $(\mathcal{H}_S\mathcal{F}, X_S^T)$ is a compressed version of the original undelayed network $(\mathcal{F}, X)$.

In this section, we show that this type of network compression can be used to gain improved stability estimates of $(\mathcal{F}, X)$. However, our main tool for gaining these improved estimates is not the time-delayed network $(\mathcal{H}_S\mathcal{F}, X_S^T)$ itself, but the undelayed version of that network.

Definition 3.12. For $S \in br_0(\Gamma_\mathcal{F})$, let $(\mathcal{R}_S\mathcal{F}, X_S)$ be the undelayed version of the time-delayed network $(\mathcal{H}_S\mathcal{F}, X_S^T)$. We call this network the *restriction* of $(\mathcal{F}, X)$ to S.

Although Lemma 3.4 states that the stability of $(\mathcal{F}, X)$ is equivalent to that of $(\mathcal{H}_S\mathcal{F}, X_S)$, there is no guarantee that the stability of $(\mathcal{R}_S\mathcal{F}, X)$ implies the stability of $(\mathcal{H}_S\mathcal{F}, X_S^T)$. To ensure that the stability of a restriction implies the stability of the original network, we define the following.

Definition 3.13. Let S be a complete branching set of the graph G. Then S is called a *basic structural set* of $\Gamma_\mathcal{F}$ if

$$|\mathcal{B}_{ij}(\Gamma_\mathcal{F}; S)| \leq 1 \text{ for all } i, j \in \mathcal{I}_S.$$

If S is a basic structural set of G, then we write $S \in st_B(G)$. The notion of a basic structural set allows for the following theorem.

Theorem 3.7. *For the network* $(\mathscr{F}, X)$, *suppose* $S \in st_B(\Gamma_{\mathscr{F}})$. *If* $\rho(\mathscr{R}_S \mathscr{F}) < 1$, *then* $(\mathscr{F}, X)$ *is stable.*

Proof. Given the dynamical network $(\mathscr{F}, X)$, suppose that $S \in st_B(\Gamma_{\mathscr{F}})$. Under this assumption, the claim is that the time-delayed system $(\mathscr{H}_S \mathscr{F}, X_S^T)$ has single-type delays. To see this, suppose $v_1, v_2 \ldots, v_j$ is a branch of $\mathscr{B}_S(\Gamma_{\mathscr{F}})$. Then in particular, the function $\mathscr{F}_{j,0}(\mathbf{x}^k) = \mathscr{F}_j(\mathbf{x}^k)$ depends on the variable x_{j-1}^k, and $\mathscr{F}_{j-1}(\mathbf{x}^k)$ on the variable x_{j-2}^k, implying that $\mathscr{F}_{j,1}(\mathbf{x}^k, \mathbf{x}^{k-1})$ is a function of x_{j-2}^{k-1}. Continuing in this manner, we conclude that

$$\mathscr{F}_{j,j-2} = \mathscr{F}_{j,j-2}(\mathbf{x}^k, \ldots, \mathbf{x}^{k-j+2})$$

is a function of x_1^{k-j+2}.

Since the index 1 is in $\mathscr{I}_S$, the component $\mathscr{H}_S \mathscr{F}_j$ must depend on the variable x_1^{k-j+2}. Therefore, $\mathscr{H}_S \mathscr{F}_j$ depends on the variable $x_i^{k-\tau+2}$ if there is a branch $v_i, \ldots, v_j \in \mathscr{B}_S(\Gamma_{\mathscr{F}})$. Conversely, if $\mathscr{H}_S \mathscr{F}_j$ depends on the variable $x_i^{k-\tau+2}$, then there must be a branch in $\mathscr{B}_S(\Gamma_{\mathscr{F}})$ from v_i to v_j. Since S is a basic structural set, the time-delayed network $(\mathscr{H}_S \mathscr{F}, X_S^T)$ has single-type delays, verifying the claim.

Assuming $\rho(\mathscr{R}_S \mathscr{F}) < 1$, then Theorem 3.4 implies that $\rho(\mathscr{H}_S \mathscr{F}) < 1$. Hence, the network $(\mathscr{H}_S \mathscr{F}, X_S^T)$ is stable by Theorem 3.1, and Lemma 3.4 then implies that $(\mathscr{F}, X)$ is stable. □

The procedure of restricting $(\mathscr{F}, X)$ can be thought of as removing the implicit delays between elements of the network indexed by elements of S. The result of this process is the lower-dimensional system $(\mathscr{R}_S \mathscr{F}, X_S)$. Theorem 3.7 states that if the restricted network is intrinsically stable, then the original "unrestricted" network is also stable. In this sense, the stability of $(\mathscr{F}, X)$ can be investigated by studying the stability of any one of its restrictions.

This procedure is illustrated in the following example.

Example 3.7. Let $(\mathscr{F}, X)$ be the dynamical network considered in Example 3.6, which has the stability matrix

$$A_{\mathscr{F}} = \begin{bmatrix} 0 & 1 & 0 & 0 & 0 & 0 \\ 1 & 0 & 1 & 0 & 0 & 0 \\ 0 & 0 & 0 & 1 & 0 & 0 \\ 0 & 0 & 0 & 0 & 1 & 0 \\ 0 & 0 & 0 & 1 & 0 & 1 \\ 1 & 0 & 0 & 0 & 0 & 0 \end{bmatrix}.$$

From this, one can compute $\rho(\mathscr{F}) = 2^{1/3}$, implying $\rho(\mathscr{F}) > 1$. That is, Theorem 3.1 cannot be used to determine whether the dynamical network $(\mathscr{F}, X)$ is stable, at least not directly.

3.4 Implicit Delays and Restrictions of Dynamical Networks

However, by removing the delays from $(\mathcal{H}_S\mathcal{F}, X_S^T)$, found in (3.24), the restriction $\mathcal{R}_S\mathcal{F}(\mathbf{x}; c)$ is given by

$$\mathcal{R}_S\mathcal{F}(\mathbf{x}; c) = \begin{bmatrix} \mathcal{R}_S\mathcal{F}_1(x_2, x_5) \\ \mathcal{R}_S\mathcal{F}_2(x_1) \\ \mathcal{R}_S\mathcal{F}_4(x_2, x_5) \\ \mathcal{R}_S\mathcal{F}_5(x_4) \end{bmatrix} = \begin{bmatrix} \tanh(\tanh(x_2) + \tanh(x_5) + c) + c \\ \tanh(x_1) + c \\ \tanh(\tanh(x_2) + \tanh(x_5) + c) + c \\ \tanh(x_4) + c \end{bmatrix}.$$

Since the branches of $\mathcal{B}_S(\Gamma_\mathcal{F})$ are given by

$$\mathcal{B}_{12}(\Gamma_\mathcal{F}; S) = v_1, v_2; \quad \mathcal{B}_{21}(\Gamma_\mathcal{F}; S) = v_2, v_1; \quad \mathcal{B}_{24}(\Gamma_\mathcal{F}; S) = v_2, v_3, v_4;$$

and

$$\mathcal{B}_{45}(\Gamma_\mathcal{F}; S) = v_4, v_5; \quad \mathcal{B}_{54}(\Gamma_\mathcal{F}; S) = v_5, v_4; \quad \mathcal{B}_{51}(\Gamma_\mathcal{F}; S) = v_5, v_6, v_1;$$

it follows that $S \in st_B(\Gamma_\mathcal{F})$.

To compute a stability matrix of the restriction $(\mathcal{R}_S\mathcal{F}, X_S)$, we note that

$$(D\mathcal{R}_S\mathcal{F})_{ij} = \text{sech}^2(x_j) \, \text{sech}^2[\tanh(x_2) + \tanh(x_5) + c] \leq$$

$$\text{sech}^2[\tanh(x_2) + \tanh(x_5) + c] \leq \text{sech}^2(c - 2)$$

for $i = 1, 4$, $j = 2, 5$, and $c \geq 2$. Similarly, $(D\mathcal{R}_S\mathcal{F})_{ij} \leq \text{sech}^2(c+2)$ for $c \leq -2$. Considering the case $c \geq 2$, we have

$$A_{\mathcal{R}_S\mathcal{F}} \leq \begin{bmatrix} 0 & 1 & 0 & 0 \\ \text{sech}^2(c-2) & 0 & \text{sech}^2(c-2) & 0 \\ 0 & 0 & 0 & 1 \\ \text{sech}^2(c-2) & 0 & \text{sech}^2(c-2) & 0 \end{bmatrix}, \quad (3.25)$$

from which it follows that

$$\rho(\mathcal{R}_S\mathcal{F}) \leq \frac{2\sqrt{2}e^2}{e^{4-c} + e^c}.$$

Thus, if $c > 2 + \ln(1 + \sqrt{2}) \approx 2.88$, then $\rho(\mathcal{R}_S\mathcal{F}) < 1$. Similarly, using the inequality $(D\mathcal{R}_S\mathcal{F})_{ij} \leq \text{sech}^2(c + 2)$ for $c \leq -2$, we have $\rho(\mathcal{R}_S\mathcal{F}) < 1$ if $c < -2 - \ln(1 + \sqrt{2})$. Therefore, Theorem 3.7 implies that $(\mathcal{F}, X)$ is stable for $|c| > 2.88$.

By restricting the network $(\mathcal{F}, X)$ to the set S, we not only change the structure of the network but also the type of weights that appears in the network's graph of interactions. This can be seen in Fig. 3.9, where $\Gamma_\mathcal{F}$ has edges with unit weight but $\Gamma_{\mathcal{R}_S\mathcal{F}}$ has weights that involve the parameter c. Since $\rho(\mathcal{F}) > 1$ and $\rho(\mathcal{R}_S\mathcal{F}) < 1$

for $|c| < 2.88$, we stress the fact that only by use of a network reduction was it possible to deduce the stability of $(\mathscr{F}, X)$ for these parameter values.

Remark 3.5. We note that a restriction $(\mathscr{R}_S\mathscr{F}, X_S)$ can be constructed without first constructing the time-delayed network $(\mathscr{H}_S\mathscr{F}, X_S)$. This can be done by letting $\mathscr{F}_{j,\ell}(\mathbf{x})$ be the function $\mathscr{F}_{j,\ell-1}(\mathbf{x})$ in which each variable $x_i \in X$ is replaced by $\mathscr{F}_i(\mathbf{x})$ if $i \notin \mathscr{I}_S$. If each variable of $\mathscr{F}_{j,m}(\mathbf{x})$ is indexed by an element of $\mathscr{I}_S$ for some $m \geq 0$, then the function $\mathscr{F}_{j,m}(\mathbf{x}_S)$ is equal to $\mathscr{R}_S\mathscr{F}_j(\mathbf{x}_S)$ for $j \in \mathscr{I}_S$.

In certain cases, it may be possible to further restrict a dynamical network and thereby obtain an improved estimate of the original network's stability.

Example 3.8. In Example 3.7, the dynamical network $(\mathscr{F}, X)$ is restricted to the basic structural set $S = \{v_1, v_2, v_4, v_5\}$. Using the restricted network $(\mathscr{R}_S\mathscr{F}, X_S)$, we were able to show that $(\mathscr{F}, X)$ is stable if $|c| > 2.88$. However, S is not the only basic structural set of this dynamical network.

In particular, it is possible to restrict $(\mathscr{F}, X)$ to the set $\tilde{S} = \{v_1, v_4\} \subset S$. Here, the result is the restriction $(\mathscr{R}_{\tilde{S}}\mathscr{F}, X_{\tilde{S}})$ given by

$$\mathscr{R}_{\tilde{S}}\mathscr{F}(\mathbf{x}; c) = \begin{bmatrix} \mathscr{R}_{\tilde{S}}\mathscr{F}_1(x_1, x_4) \\ \mathscr{R}_{\tilde{S}}\mathscr{F}_4(x_1, x_4) \end{bmatrix}$$

$$= \begin{bmatrix} \tanh[\tanh(\tanh(x_1) + c) + \tanh(\tanh(x_4)) + c] + c \\ \tanh[\tanh(\tanh(x_1) + c) + \tanh(\tanh(x_4)) + c] + c \end{bmatrix}.$$

For $c = 1$, one can show that the stability matrix of $(\mathscr{R}_{\tilde{S}}\mathscr{F}, X_{\tilde{S}})$ is the matrix

$$A_{\mathscr{R}_{\tilde{S}}\mathscr{F}} = \begin{bmatrix} 0.482 & 0.482 \\ 0.482 & 0.482 \end{bmatrix}.$$

Since $\rho(\mathscr{R}_{\tilde{S}}\mathscr{F}) = 0.964$, Theorem 3.7 implies that $(\mathscr{F}, X)$ is stable at this value of c.

If one computes the stability matrix of $(\mathscr{R}_S\mathscr{F}, X_S)$ for $c = 1$, the result is

$$A_{\mathscr{R}_S\mathscr{F}} = \begin{bmatrix} 0 & 1 & 0 & 0 \\ 1 & 0 & 1 & 0 \\ 0 & 0 & 0 & 1 \\ 1 & 0 & 1 & 0 \end{bmatrix}, \tag{3.26}$$

for which $\rho(\mathscr{R}_S\mathscr{F}) = \sqrt{2} > 1$. For this parameter value, the restriction of $(\mathscr{F}, X)$ to the structural set S cannot be used to establish the stability of the original unreduced network. It is only by restricting $(\mathscr{F}, X)$ to the smaller structural set $\tilde{S} \subset S$ that we are able to show that this network is stable at the parameter value $c = 1$.

3.4 Implicit Delays and Restrictions of Dynamical Networks

We note that by combining Theorems 3.4 and 3.7, it is possible to use network restrictions to get improved stability estimates of time-delayed networks.

Corollary 3.4. *Suppose* $(\mathcal{H}, X^T)$ *has single-type delays. If* $(\mathcal{U}, X)$ *has the restriction* $(\mathcal{R}_S \mathcal{U}, X_S)$ *with* $S \in st_B(\Gamma_\mathcal{U})$ *and spectral radius* $\rho(\mathcal{R}_S \mathcal{U}) < 1$, *then* $(\mathcal{H}, X^T)$ *is stable.*

At this point, we have described how to take a dynamical network $(\mathcal{F}, X)$, find its time-delayed version $(\mathcal{H}_S \mathcal{F}, X_S^T)$ with respect to some $S \in br_0(\Gamma_\mathcal{F})$, and then remove the system's delays to create the restriction $(\mathcal{R}_S \mathcal{F}, X_S)$. This last step of removing a system's time delays has the advantage that it simplifies the dynamical network both in its dynamics and in how it is formally represented.

However, by removing the system's delays, we end up modifying both the system's dynamics and typically its entire spectrum. If want to preserve the network's dynamics and spectrum to a large degree, then there is another option. Rather than removing the delays of $(\mathcal{H}_S \mathcal{F}, X_S^T)$, we simply consider the dynamical network $(\overline{\mathcal{H}_S \mathcal{F}}, X_S^T)$, which is the undelayed version of the system (see Definition 3.4).

Definition 3.14. For the dynamical network $(\mathcal{F}, X)$, let $S \in br_0(\Gamma_\mathcal{F})$. The dynamical network $(\mathcal{X}_S \mathcal{F}, X_S^T) = (\overline{\mathcal{H}_S \mathcal{F}}, X_S^T)$ is called the *dynamical network expansion* of $(\mathcal{F}, X)$ with respect to S.

By combining Lemmas 3.1 and 3.4 with Theorem 3.1, we have the following theorem, relating the stability of a dynamical network and its expansions.

Theorem 3.8. *Let* $(\mathcal{X}_S \mathcal{F}, X_S^T)$ *be a dynamical network expansion of* $(\mathcal{F}, X)$. *Then* $(\mathcal{F}, X)$ *is stable if* $\rho(\mathcal{X}_S \mathcal{F}) < 1$.

Example 3.9. Consider the dynamical network $(\mathcal{F}, X)$, similar to that in Example 3.6, given by

$$\mathcal{F}(\mathbf{x}; c) = \begin{bmatrix} \mathcal{F}_1(x_1, x_2, x_4) \\ \mathcal{F}_2(x_1) \\ \mathcal{F}_3(x_2, x_3, x_4) \\ \mathcal{F}_4(x_3) \end{bmatrix} = \begin{bmatrix} \frac{1}{2}\tanh(x_1) + \tanh(x_2) + \tanh(x_4) + c \\ \tanh(x_1) + c \\ \tanh(x_2) + \frac{1}{2}\tanh(x_3) + \tanh(x_4) + c \\ \tanh(x_3) + c \end{bmatrix},$$

where $\mathbf{x} \in \mathbb{R}^4$ and $c \in \mathbb{R}$. Here, one can compute that the dynamical network has the stability matrix

$$A_\mathcal{F} = \begin{bmatrix} \frac{1}{2} & 1 & 0 & 1 \\ 1 & 0 & 0 & 0 \\ 0 & 1 & \frac{1}{2} & 1 \\ 0 & 0 & 1 & 0 \end{bmatrix} \quad \text{with spectral radius } \rho(A_\mathcal{F}) = \frac{1}{4}(1 + \sqrt{33}). \tag{3.27}$$

Since $\rho(A_\mathcal{F}) > 1$, we are again in the situation in which the stability of the dynamical network cannot be determined by direct use of Theorem 3.1. Additionally, as can be seen from Fig. 3.10, the graph of interactions $\Gamma_\mathcal{F}$ has no nontrivial basic

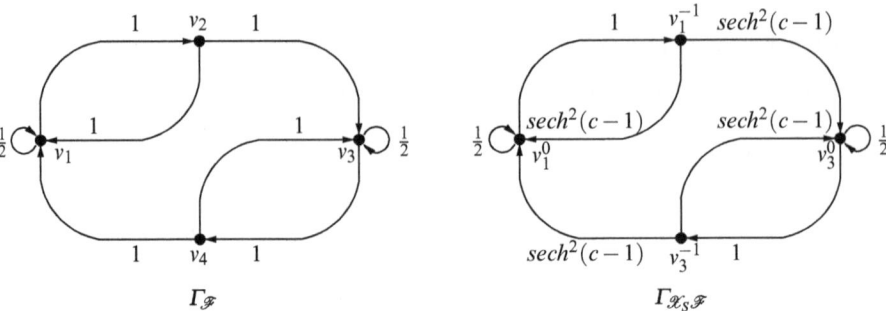

Fig. 3.10 The graph of interactions $\Gamma_{\mathcal{F}}$ of the network in Example 3.9 and the graph of its expansion $\Gamma_{\mathcal{X}_S\mathcal{F}}$ over the complete branching set $S = \{v_1, v_3\}$

structural sets, so there is no way to use a network restriction to determine the stability of $(\mathcal{F}, X)$.

The set of vertices $S = \{v_1, v_3\}$, however, is a complete branching set of $\Gamma_{\mathcal{F}}$. Therefore, we can construct the time-delayed version of this dynamical network $(\mathcal{H}_S\mathcal{F}, X_S^T)$, which is

$$\mathbf{x}^{k+1} = \begin{bmatrix} \frac{1}{2}\tanh(x_1^k) + \tanh(\tanh(x_1^{k-1}) + c) + \tanh(\tanh(x_3^{k-1}) + c) + c \\ \tanh(\tanh(x_1^{k-1}) + c) + \frac{1}{2}\tanh(x_3^k) + \tanh(\tanh(x_3^{k-1}) + c) + c \end{bmatrix},$$

where $T = 2$. The undelayed version of $(\mathcal{H}_S\mathcal{F}, X_S^T)$, which is the dynamical network expansion $(\mathcal{X}_S\mathcal{F}, X_S^T)$, is given by

$$\mathcal{X}_S\mathcal{F}(\mathbf{x}; c) = \begin{bmatrix} \mathcal{X}_S\mathcal{F}_1^0(x_1^0, x_1^{-1}, x_3^{-1}) \\ \mathcal{X}_S\mathcal{F}_1^{-1}(x_1^0) \\ \mathcal{X}_S\mathcal{F}_2^0(x_1^{-1}, x_3^0, x_3^{-1}) \\ \mathcal{X}_S\mathcal{F}_2^{-1}(x_3^0) \end{bmatrix}$$

$$= \begin{bmatrix} \frac{1}{2}\tanh(x_1^0) + \tanh(\tanh(x_1^{-1}) + c) + \tanh(\tanh(x_3^{-1}) + c) + c \\ x_1^0 \\ \tanh(\tanh(x_1^{-1}) + c) + \frac{1}{2}\tanh(x_3^0) + \tanh(\tanh(x_3^{-1}) + c) + c \\ x_3^0 \end{bmatrix}.$$

Using the same type of argument as in Example 3.8 to bound the entries of $A_{\mathcal{X}_S\mathcal{F}}$, we find that

$$A_{\mathcal{X}_S\mathcal{F}} \leq \begin{bmatrix} \frac{1}{2}\operatorname{sech}^2(c-1) & 0 & \operatorname{sech}^2(c-1) \\ 1 & 0 & 0 & 0 \\ 0 & \operatorname{sech}^2(c-1) & \frac{1}{2}\operatorname{sech}^2(c-1) \\ 0 & 0 & 1 & 0 \end{bmatrix} \quad (3.28)$$

3.4 Implicit Delays and Restrictions of Dynamical Networks

for $c > 1$. From this, it follows that

$$\rho(\mathscr{X}_S\mathscr{F}) \leq \frac{e^2 + e^{2c} + e^c\sqrt{34e^2 + e^{4-2c}} + e^{2c}}{4(e^2 + e^{2c})},$$

so that if $c > \log[(2 + \sqrt{3})e] \approx 2.31$, then $\rho(\mathscr{X}_S\mathscr{F}) < 1$. Similarly, one can show that if $c < -\log[(2 + \sqrt{3})e] \approx -2.31$, then $\rho(\mathscr{X}_S\mathscr{F}) < 1$, so that Theorem 3.8 implies that $(\mathscr{F}, X)$ is globally stable if $|c| > 2.31$.

Since every basic structural set is a complete branching set, dynamical network expansions are a more general tool than network restrictions for determining the stability of a given dynamical network. The main reason we consider both is that network restrictions are much easier to construct and analyze than dynamical network expansions. In fact, if S is a basic structural set of $\Gamma_\mathscr{F}$, then $(\mathscr{R}_S\mathscr{F}, X_S)$ is always of lower dimension than $(\mathscr{X}_S\mathscr{F}, X_S^T)$.

Dynamical network expansions and restrictions are similar, though, in the sense that they are built around the same idea. If we are going to use a stability matrix, i.e., a global linearization, to evaluate the stability of a dynamical network, then the more we can compress the network around a specific set of network elements, the better our estimates become. These specific elements are, of course, the structural sets over which we choose to restrict or expand our network.

We note that if we have a linear network, i.e., a network $(\mathscr{F}, X)$, where $\mathscr{F} : X \to X$ is described by a matrix, then the processes of restricting and expanding a network do not change the network's spectral radius. In a nonlinear network, or more generally dynamical system, the operations of restricting and expanding the system lead to an averaging of the network's nonlinear processes, which allows for better estimates of the network's overall behavior.

Chapter 4
Improved Eigenvalue Estimates

The classical Abel–Ruffini theorem, often called Abel's impossibility theorem, states that there is no algebraic formula for the roots of the general polynomial of degree five or higher. Consequently, one cannot, in general, compute the spectrum of a matrix $A \in \mathbb{C}^{n \times n}$, for $n \geq 5$.

Beginning on the mid-nineteenth century, a number of methods were developed to approximate the eigenvalues of complex-valued matrices. The main idea in each of these methods is to associate a bounded region of the complex plane to each $A \in \mathbb{C}^{n \times n}$ that contains $\sigma(A)$. This bounded region is the approximation of the eigenvalues of A.

In this chapter, we investigate how isospectral matrix reductions affect these approximation methods. Our main result is that isospectral reductions can be used in conjunction with each of these methods to gain improved eigenvalue estimates of any matrix $A \in \mathbb{C}^{n \times n}$. Specifically, we show that under certain conditions, the eigenvalue region associated with a reduced matrix is contained in the eigenvalue region associated with the original unreduced matrix.

Informally, the way this works is the following. The classical methods we consider associate to each $A \in \mathbb{C}^{n \times n}$ an eigenvalue region in the complex plane, which is, in fact, the union of a number of subregions. Each of these subregions is computed based on a limited number of rows of A. In this sense, these subregions, and therefore the eigenvalue approximations, use only "local information" from the matrix to approximate its eigenvalues.

By isospectrally reducing a matrix, we are in a sense compressing the information contained in the matrix. Hence, the local information in the smaller reduced matrix comes from information in the larger matrix that is less local. This increase in local information in the reduced matrix allows us, via these classical methods, to gain improved eigenvalue estimates.

We show, in two of the three classical methods we consider, that isospectral matrix reductions always lead to improved eigenvalue estimates. For the third method, we give a sufficient condition under which an isospectral reduction leads to improved estimates.

We apply these techniques in a variety of settings, including estimating the spectra of combinatorial and normalized graph Laplacians. Additionally, we demonstrate how this approach can be used to estimate the eigenvalues of large matrices or graphs in which only limited or local information is known. We also apply these techniques to the problem of estimating the spectral radius of a graph (network), which has implications to the stability of dynamical networks.

4.1 Gershgorin-Type Regions

A theorem of Gershgorin's, originating from [21], gives a simple method for bounding the eigenvalues of a square matrix with complex-valued entries. This result is found in Theorem 4.1, which we formulate after introducing some notation.

If $A \in \mathbb{C}^{n \times n}$, let

$$r_i(A) = \sum_{j=1,\ j \neq i}^{n} |A_{ij}|, \quad 1 \leq i \leq n, \tag{4.1}$$

be the *i*th *absolute row sum* of A.

Theorem 4.1 (Gershgorin [21]). *Let $A \in \mathbb{C}^{n \times n}$. Then all eigenvalues of A are contained in the set*

$$\Gamma(A) = \bigcup_{i=1}^{n} \{\lambda \in \mathbb{C} : |\lambda - A_{ii}| \leq r_i(A)\}.$$

Geometrically, Gershgorin's theorem states that the eigenvalues of $A \in \mathbb{C}^{n \times n}$ are contained in the union of n circles in the complex plane. The *i*th circle, corresponding to the *i*th row of A, is centered at the diagonal entry $A_{ii} \in \mathbb{C}$ with radius $r_i(A)$.

Example 4.1. Let $A \in \mathbb{C}^{3 \times 3}$ be the matrix with entries

$$A = \begin{bmatrix} -3 & 0 & 1 \\ 1 & 1 & -1 \\ 1 & 0 & 5 \end{bmatrix},$$

which has the eigenvalues $\sigma(A) = \{1, 1 \pm \sqrt{17}\}$. The region $\Gamma(A)$ is made up of the three circles shown in Fig. 4.1 (left). The blue, red, and tan circles correspond to the first, second, and third rows of A, respectively. The spectrum of A is also indicated.

Recall from Chap. 1 that if a matrix $M(\lambda)$ is in $\mathbb{W}^{n \times n}$, then this matrix has a well-defined spectrum. A natural question is whether Gershgorin's theorem can be

4.1 Gershgorin-Type Regions

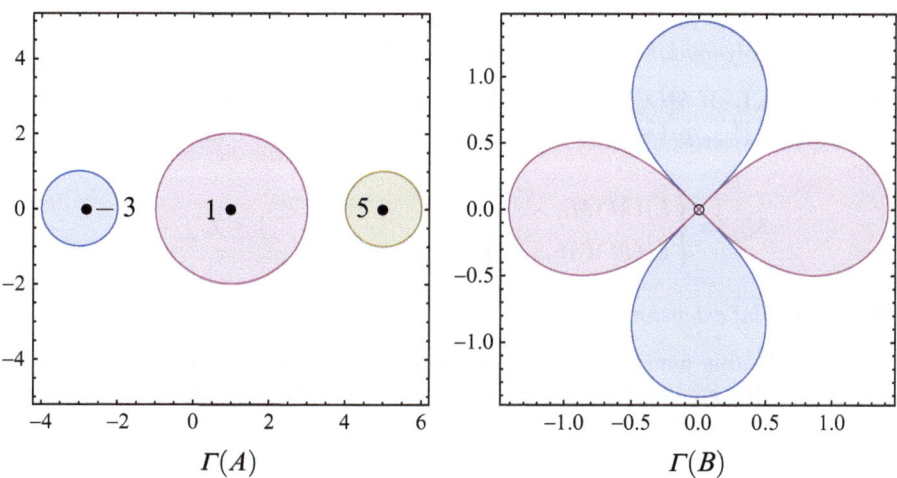

Fig. 4.1 The Gershgorin regions for the matrices $A \in \mathbb{C}^{3 \times 3}$ (*left*) and $B(\lambda) \in \mathbb{W}^{2 \times 2}$ (*right*) in Examples 4.1 and 4.2, respectively

directly applied to $M(\lambda)$ to estimate its spectrum as if it were a matrix with complex entries. This we consider in the following example.

Example 4.2. Consider the matrix $B(\lambda) \in \mathbb{W}^{2 \times 2}$ given by

$$B(\lambda) = \begin{bmatrix} -\frac{1}{\lambda} & -\frac{1}{\lambda} \\ \frac{1}{\lambda} & \frac{1}{\lambda} \end{bmatrix}.$$

In principle, it is possible to use Gershgorin's theorem formally on the matrix $B(\lambda)$. The resulting region

$$\Gamma(B) = \left\{ \lambda \in \mathbb{C} : \left|\lambda + \frac{1}{\lambda}\right| \leq \left|\frac{1}{\lambda}\right| \right\} \cup \left\{ \lambda \in \mathbb{C} : \left|\lambda - \frac{1}{\lambda}\right| \leq \left|\frac{1}{\lambda}\right| \right\}$$

is shown in Fig. 4.1 (right). As one can compute, $\det(B(\lambda) - \lambda I) = \lambda^2$, implying $\sigma(B) = \{0, 0\}$. However, the region $\Gamma(B)$ does not include the eigenvalue(s) $0 \in \sigma(B)$. This is indicated by the open circle in the figure.

Based on Example 4.2, we conclude that Gershgorin's theorem does not directly apply to matrices with rational function entries. However, if we take a closer look at Example 4.2, we note that the reason we have $\sigma(B) \nsubseteq \Gamma(B)$ is because $\sigma(B) \nsubseteq dom(B)$.

Since the Gershgorin region $\Gamma(M)$ of every $M \in \mathbb{W}^{n \times n}$ is contained in $dom(M)$, only the eigenvalues of M in its domain can be in its Gershgorin region. To adapt Gershgorin's theorem to the matrices $\mathbb{W}^{n \times n}$, we need a way of including those eigenvalues of M that are not in its domain. With this in mind, we define the notion of a *polynomial extension* of a matrix $M \in \mathbb{W}^{n \times n}$.

Definition 4.1. If $M(\lambda) \in \mathbb{W}^{n \times n}$ and $M_{ij} = p_{ij}/q_{ij}$, where $p_{ij}(\lambda), q_{ij}(\lambda) \in \mathbb{C}[\lambda]$, let $L_i(M) = \prod_{j=1}^{n} q_{ij}(\lambda)$ for $1 \leq i \leq n$. We call the matrix $\bar{M}$ given by

$$\bar{M}_{ij} = \begin{cases} L_i(M) M_{ij} & i \neq j, \\ L_i(M)(M_{ij} - \lambda) + \lambda, & i = j \end{cases} \quad 1 \leq i, j \leq n,$$

the *polynomial extension* of M.

To justify this name, note that each entry $\bar{M}_{ij}$ is an element of $\mathbb{C}[\lambda]$, or $\bar{M}$ has entries that are complex polynomials. The reason we consider the polynomial extension of a matrix $M \in \mathbb{W}^{n \times n}$ is that $dom(\bar{M}) = \mathbb{C}$, so that $\bar{M}(\lambda)$ is defined everywhere. Moreover, the spectrum of M is contained in the spectrum of $\bar{M}$.

Lemma 4.1. *Let $M(\lambda) \in \mathbb{W}^{n \times n}$. Then $\sigma(M) \subseteq \sigma(\bar{M})$.*

Proof. For $M \in \mathbb{W}^{n \times n}$, note that the matrix $\bar{M} - \lambda I$ is given by

$$(\bar{M} - \lambda I)_{ij} = \begin{cases} L_i(M) M_{ij} & i \neq j, \\ L_i(M)(M_{ij} - \lambda), & i = j \end{cases} \quad \text{for } 1 \leq i \leq n.$$

The matrix $\bar{M} - \lambda I$ is then the matrix $M - \lambda I$ whose ith row has been multiplied by L_i. Therefore,

$$\det(\bar{M} - \lambda I) = \left(\prod_{i=1}^{n} L_i(M) \right) \det(M - \lambda I),$$

implying $\sigma(M) \subseteq \sigma(\bar{M})$. □

In order to extend Theorem 4.1 to the matrices $\mathbb{W}^{n \times n}$, we use the following adaptation of the notation given by (4.1). For $M \in \mathbb{W}^{n \times n}$, let

$$r_i(M) = \sum_{j=1, j \neq i}^{n} |M_{ij}| \text{ for } 1 \leq i \leq n$$

be the *ith absolute row sum* of M.

Note that since $\bar{M} \in \mathbb{C}[\lambda]^{n \times n}$ for every $M \in \mathbb{W}^{n \times n}$, we can view $\bar{M}(\lambda)$ as a function $\bar{M}(\cdot) : \mathbb{C} \to \mathbb{C}^{n \times n}$ and $\bar{M}(\cdot)_{ij} : \mathbb{C} \to \mathbb{C}$. Likewise, we can consider $r_i(\bar{M}) = r_i(\bar{M}, \lambda)$ to be the function $r_i(\bar{M}, \cdot) : \mathbb{C} \to \mathbb{C}$. However, we will typically suppress the dependence of $\bar{M}$ and $r_i(\bar{M})$ on λ, for ease of notation.

Theorem 4.2. *Let $M(\lambda) \in \mathbb{W}^{n \times n}$. Then $\sigma(M)$ is contained in the set*

$$\Gamma_{\mathbb{W}}(M) = \bigcup_{i=1}^{n} \{\lambda \in \mathbb{C} : |\lambda - \bar{M}_{ii}| \leq r_i(\bar{M})\}.$$

4.1 Gershgorin-Type Regions

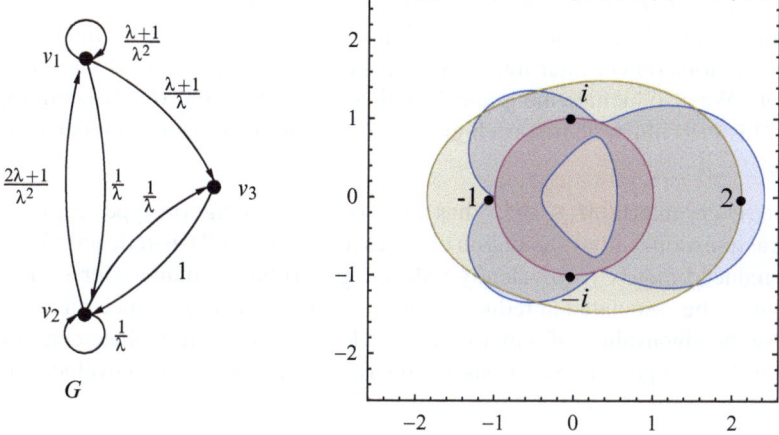

Fig. 4.2 For M in Example 4.3, the graph G, where $M(G) = M$, is shown (*left*) and the region $\Gamma_\mathbb{W}(M)$ is shown (*right*), where $\sigma(M) = \{-1, -1, 2, -i, i\}$ is indicated

Proof. First note that for $\alpha \in \sigma(M)$, the matrix $\bar{M}(\alpha)$ is in $\mathbb{C}^{n \times n}$. Since Lemma 4.1 implies that α is an eigenvalue of $\bar{M}(\alpha)$, then by an application of Gershgorin's theorem, the inequality $|\alpha - \bar{M}(\alpha)_{ii}| \leq r_i(\bar{M}, \alpha)$ holds for some $1 \leq i \leq n$. Hence, $\alpha \in \Gamma_\mathbb{W}(M)$. □

Because it will be useful later in comparing different regions in the complex plane, for $M \in \mathbb{W}^{n \times n}$ we let

$$\Gamma_\mathbb{W}(M)_i = \{\lambda \in \mathbb{C} : |\lambda - \bar{M}_{ii}| \leq r_i(\bar{M})\}, \text{ where } 1 \leq i \leq n,$$

and call this the *ith Gershgorin-type region* of M. Similarly, we call the union $\Gamma_\mathbb{W}(M)$ of these n sets the *Gershgorin-type region* of the matrix M.

As an illustration of Theorem 4.2, consider the following example.

Example 4.3. Let $M \in \mathbb{W}^{n \times n}$ be the matrix given by

$$M = \begin{bmatrix} \frac{\lambda+1}{\lambda^2} & \frac{1}{\lambda} & \frac{\lambda+1}{\lambda} \\ \frac{2\lambda+1}{\lambda^2} & \frac{1}{\lambda} & \frac{1}{\lambda} \\ 0 & 1 & 0 \end{bmatrix}. \tag{4.2}$$

Since $\det(M(\lambda) - \lambda I) = (-\lambda^5 + 2\lambda^3 + 2\lambda^2 + 3\lambda + 2)/(\lambda^2)$, one can compute that $\sigma(M) = \{-1, -1, i, -i, 2\}$. The corresponding Gershgorin-type region $\Gamma_\mathbb{W}(M)$ is shown in Fig. 4.2 (right), where

$$\bar{M} = \begin{bmatrix} -\lambda^5 + \lambda^3 + \lambda^2 + \lambda & \lambda^3 & \lambda^4 + \lambda^3 \\ 2\lambda^3 + \lambda^2 & -\lambda^5 + \lambda^3 + \lambda & \lambda^3 \\ 0 & 1 & 0 \end{bmatrix}.$$

The set $\Gamma_{\mathbb{W}}(M)$ is the union of the three regions $\Gamma_{\mathbb{W}}(M)_1$, $\Gamma_{\mathbb{W}}(M)_2$, and $\Gamma_{\mathbb{W}}(M)_3$, whose boundaries are shown in blue, red, and tan. The interior colors of these regions reflect their intersections, and the eigenvalues $\sigma(M)$ are indicated as points. We also include the graph G with adjacency matrix M, shown in Fig. 4.2 (left). This is to illustrate that each graph $G \in \mathbb{G}$ has an associated Gershgorin-type region.

Since every matrix $M \in \mathbb{W}_\pi^{n \times n}$ has an associated Gershgorin-type region $\Gamma(M)$, a natural question is how this region changes as the matrix M is reduced. As it turns out, a reduced matrix (equivalently reduced graph) has a smaller Gershgorin-type region than the associated unreduced matrix (graph). Because isospectral reductions preserve the eigenvalues of a matrix, up to a known set, an immediate consequence of this is that isospectral reductions can be used to improve the eigenvalue estimates given in Theorems 4.1 and 4.2.

Theorem 4.3 (Improved Gershgorin Regions). *For $M(\lambda) \in \mathbb{W}_\pi^{n \times n}$, suppose that the set $S \subset N$ is nonempty. Then $\Gamma_{\mathbb{W}}(\mathscr{R}(M;S)) \subseteq \Gamma_{\mathbb{W}}(M)$.*

In terms of graphs, Gershgorin's original theorem can be thought of as estimating the spectrum of a graph G with $M(G) \in \mathbb{C}^{n \times n}$ by considering the paths of length 1 starting at each graph vertex. Isospectral graph reductions allow for better eigenvalue estimates by considering longer paths in the graph through those vertices that have been removed. That is, an isospectral graph reduction collapses the information stored in a branch of G to a single edge of $\mathscr{R}_S(G)$. The entire branch, and therefore more information, is then considered in calculating $\Gamma_{\mathbb{W}}(\mathscr{R}(M;S))$, which leads to better eigenvalue estimates.

Theorem 4.3 together with Theorem 1.1 has the following corollary.

Corollary 4.1. *For $M(\lambda) \in \mathbb{W}_\pi^{n \times n}$, suppose that the set $S \subset N$ is nonempty. Then*

$$\sigma(M) \subseteq \Gamma_{\mathbb{W}}(\mathscr{R}(M;S)) \cup \sigma(M|_{\bar{S}\bar{S}}).$$

Proof. Since $\Gamma_{\mathbb{W}}(\mathscr{R}(M;S)) \subseteq \Gamma_{\mathbb{W}}(M)$, it follows from Theorem 1.1 that

$$\sigma(M) \subseteq \Gamma_{\mathbb{W}}(\mathscr{R}(M;S)) \cup \sigma(M|_{\bar{S}\bar{S}}) \cup \sigma^{-1}(M).$$

Since $\det(M(\lambda) - \lambda I) = p(\lambda)/q(\lambda)$ for some $p(\lambda), q(\lambda) \in \mathbb{C}[\lambda]$ with no common factors, we have $\sigma(M) \cap \sigma^{-1}(M) = \emptyset$, and the result follows. □

In order to prove Theorem 4.3, we will need to be able to evaluate functions in $\mathbb{W}$ at some fixed value $\lambda \in \mathbb{C}$. In each case, we consider such functions first as elements in $\mathbb{W}$ with common factors removed, which we will then evaluate at λ. In fact, most of these functions, once common factors are removed, will be polynomials in $\mathbb{C}[\lambda]$.

To simplify notation, we will use the following. For $M(\lambda) \in \mathbb{W}^{n \times n}$, where $n \geq 2$ and $S = \{2, \ldots, n\}$, we let $\mathscr{R}(M;S) = \mathscr{R}_1$, $L_k(M) = L_k$, $L_k(\mathscr{R}_1) = L_k^1$, $M_{k\ell} = \omega_{k\ell}$, and $\lambda - \omega_{kk} = \lambda_{kk}$. Also, we let $\omega_{k\ell} = p_{k\ell}/q_{k\ell}$ for $p_{k\ell}, q_{k\ell} \in \mathbb{C}[\lambda]$, where we assume that $q_{k\ell} = 1$ if $\omega_{k\ell} = 0$. Lastly, we set $R_k(M) = \sum_{\ell=1, \ell \neq k}^{n} |\omega_{k\ell} L_k|$.

4.1 Gershgorin-Type Regions

Before proceeding, we state the following lemma.

Lemma 4.2. *If $M(\lambda) \in \mathbb{W}_\pi^{n \times n}$ for $n \geq 2$, then $q_{11} q_{i1} L_i^1 = \big(q_{i1}(q_{11}\lambda - p_{11})\big)^{n-1} L_1 L_i$.*

Proof. First, note that

$$(\mathscr{R}_1)_{ij} = \frac{p_{i1} p_{1j} q_{ij} q_{11} + q_{i1} q_{1j} p_{ij} (q_{11}\lambda - p_{11})}{q_{i1} q_{1j} q_{ij} (q_{11}\lambda - p_{11})} \quad \text{for each } 2 \leq i, j \leq n,$$

from which it follows that $L_i^1 = \prod_{j=2}^{n} q_{i1} q_{1j} q_{ij} (q_{11}\lambda - p_{11})$. Therefore,

$$L_i^1 = \big(q_{i1}(q_{11}\lambda - p_{11})\big)^{n-1} \prod_{j=2}^{n} q_{1j} \prod_{j=2}^{n} q_{ij}. \tag{4.3}$$

Since $L_k = \prod_{j=1}^{n} q_{kj}$ for $1 \leq k \leq n$, the result follows by multiplication of $q_{11} q_{i1}$ to both sides of this equation. $\square$

A proof of Theorem 4.3 is the following.

Proof. Suppose that $\lambda \in \Gamma_\mathbb{W}(\mathscr{R}_1)_i$ for fixed $\lambda \in \mathbb{C}$ and $2 \leq i \leq n$. Since each $(\mathscr{R}_1)_{ij}$ is equal to $\omega_{ij} + \omega_{i1} \omega_{1j} / \lambda_{11}$ for $2 \leq j \leq n$, we have

$$\left|\left(\lambda_{ii} - \frac{\omega_{i1} \omega_{1i}}{\lambda_{11}}\right) L_i^1\right| \leq \sum_{j=2, j \neq i}^{n} \left|\left(\omega_{ij} + \frac{\omega_{i1} \omega_{1j}}{\lambda_{11}}\right) L_i^1\right|.$$

Multiplying both sides of this inequality by $|\lambda_{11} q_{11} q_{i1}|$ implies, via Lemma 4.2, that

$$Q_i(M) |\lambda_{11} L_1 \lambda_{ii} L_i - \omega_{i1} \omega_{1i} L_1 L_i| \leq Q_i(M) \sum_{j=2, j \neq i}^{n} |(\omega_{ij} \lambda_{11} + \omega_{i1} \omega_{1j}) L_1 L_i|,$$

where $Q_i(M) = |\big(q_{i1}(q_{11}\lambda - p_{11})\big)|^{n-1}$. If $Q_i(M) \neq 0$, then by the triangle inequality,

$$|\lambda_{11} L_1 \lambda_{ii} L_i| - |\omega_{i1} \omega_{1i} L_1 L_i| \leq \sum_{j=2, j \neq i}^{n} |\lambda_{11} L_1 \omega_{ij} L_i| + \sum_{j=2, j \neq i}^{n} |\omega_{i1} L_i \omega_{1j} L_1|.$$

Therefore,

$$|\lambda_{11}L_1\lambda_{ii}L_i| - \sum_{j=1,j\neq i}^{n}|\lambda_{11}L_1\omega_{ij}L_i| \leq \sum_{j=2}^{n}|\omega_{i1}\omega_{1j}L_1L_i| - |\omega_{i1}L_i\lambda_{11}L_1|.$$

By factoring, we have

$$|\lambda_{11}L_1|\bigl(|\lambda_{ii}L_i| - R_i(M)\bigr) \leq |\omega_{i1}L_i|\bigl(R_1(M) - |\lambda_{11}L_1|\bigr). \tag{4.4}$$

If we assume $\lambda \notin \Gamma_W(G)_i \cup \Gamma_W(G)_1$, then we have both

$$|\lambda_{ii}L_i| - R_i(M) > 0 \text{ and } R_1(M) - |\lambda_{11}L_1| < 0.$$

These inequalities together with (4.4) imply that $\lambda_{11}L_1 = 0$. However, this, in turn, implies that $\lambda \in \Gamma_W(M)_1$, which is impossible.

Hence, $\lambda \in \Gamma_W(M)_i \cup \Gamma_W(M)_1$ unless $Q_i(M) = 0$. Supposing, then, that this is the case, note that if $L_{ij} = \prod_{\ell=1,\ell\neq j}^{n} q_{i\ell}$ for $1 \leq i,j \leq n$, then

$$\Gamma_W(M)_k = \{\lambda \in \mathbb{C} : |L_{kk}(q_{kk}\lambda - p_{kk})| \leq \sum_{j=1,j\neq k}^{n}|p_{kj}L_{kj}|\} \text{ for } 1 \leq k \leq n. \tag{4.5}$$

Under the assumption $Q_i(M) = \bigl(q_{i1}(q_{11}\lambda - p_{11})\bigr)^{n-1} = 0$, note that if $q_{i1} = 0$, then $L_{ii} = 0$, implying $\lambda \in \Gamma_W(M)_i$. If $q_{11}\lambda - p_{11} = 0$, then $\lambda \in \Gamma_W(M)_1$ by (4.5).

It then follows that

$$\Gamma_W(\mathcal{R}_1)_i \subseteq \Gamma_W(M)_1 \cup \Gamma_W(M)_i, \tag{4.6}$$

implying $\Gamma_W(\mathcal{R}_1) \subseteq \Gamma_W(M)$. The theorem follows by repeated use of Theorem 1.3, since it is always possible to reduce the matrix M sequentially over an arbitrary index set S by removing a single index at each step. □

The reason Gershgorin-type estimates improve under reduction can be found in equation (4.6). From the point of view of graph reduction, the inclusion given in (4.6) states that if we remove the vertex v_j from a graph, then the ith Gershgorin region associated with the reduced graph is contained in the union of the ith and jth Gershgorin regions of the unreduced graph.

In Fig. 4.3 (left), we show the inclusion

$$\Gamma_W(\mathcal{R}(M;S))_2 \subset \Gamma_W(M)_1 \cup \Gamma_W(M)_2$$

4.1 Gershgorin-Type Regions

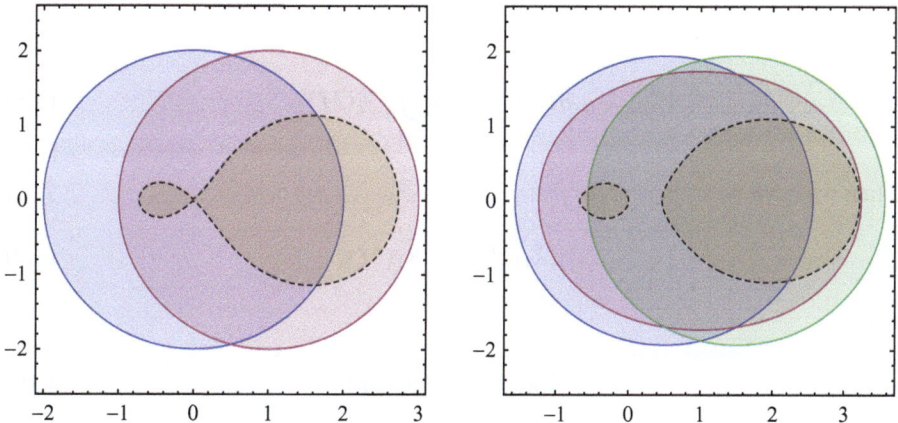

Fig. 4.3 *Left*: the regions $\Gamma_{\mathbb{W}}(M)_1$ (blue), $\Gamma_{\mathbb{W}}(M)_2$ (red), and $\Gamma_{\mathbb{W}}(\mathscr{R}(M;S))_2$ (tan, with dashed boundary) are shown. *Right*: the regions $K_{\mathbb{W}}(M)_{12}$ (*blue*), $K_{\mathbb{W}}(M)_{13}$ (*red*), $K_{\mathbb{W}}(M)_{23}$ (*green*), and $K_{\mathbb{W}}(\mathscr{R}(M;S))_{23}$ (tan, with *dashed boundary*) are shown

for the matrix $M \in \mathbb{C}^{3 \times 3}$ given by

$$M = \begin{bmatrix} 0 & 1 & 1 \\ 1 & 1 & 1 \\ 1 & 1 & 2 \end{bmatrix}, \tag{4.7}$$

where the index set S is equal to $\{2, 3\}$.

To understand in which situations $\Gamma_{\mathbb{W}}(\mathscr{R}(M;S))$ is strictly contained in $\Gamma_{\mathbb{W}}(M)$, we consider the following. For $M \in \mathbb{W}_\pi^{n \times n}$, let

$$\partial \Gamma_{\mathbb{W}}(M)_i = \{\lambda \in \mathbb{C} : |\lambda - \bar{M}_{ii}| = r_i(\bar{M})\} \text{ for } 1 \leq i \leq n.$$

We note here that the boundary of the region $\Gamma_{\mathbb{W}}(M)_i$ in the complex plane is contained in the set $\partial \Gamma_{\mathbb{W}}(M)_i$ for each $1 \leq i \leq n$. This follows from the continuity of $|\lambda - \bar{M}_{ii}| - r_i(\bar{M})$ in the variable λ. However, if $\lambda \in \partial \Gamma_{\mathbb{W}}(M)_i$, it may be the case that λ is contained in a neighborhood lying entirely within $\Gamma_{\mathbb{W}}(M)_i$, in which case λ is not on the boundary of $\Gamma_{\mathbb{W}}(M)_i$. That is, the boundary of the set $\Gamma_{\mathbb{W}}(M)_i$ is contained in what we have defined as $\partial \Gamma_{\mathbb{W}}(M)_i$, but this containment may not be strict.

Theorem 4.4. *Let $M(\lambda) \in \mathbb{W}_\pi^{n \times n}$. Suppose the subset*

$$\partial \Gamma_{\mathbb{W}}(M)_i \setminus \bigcup_{j=1, j \neq i}^{n} \Gamma_{\mathbb{W}}(M)_j$$

is an infinite set of points. Then $\Gamma_{\mathbb{W}}(\mathscr{R}(M;S)) \subset \Gamma_{\mathbb{W}}(M)$ for every nonempty $S \subset N$ if $i \notin S$.

Proof. Let $\lambda \in \mathbb{C}$ be fixed such that

$$\lambda \in \partial \Gamma_{\mathbb{W}}(M)_1 \setminus \bigcup_{j=2}^{n} \Gamma_{\mathbb{W}}(M)_j. \tag{4.8}$$

Then we have both

$$|(\lambda_{11})L_1| = R_1(M), \tag{4.9}$$

and

$$|(\lambda_{ii})L_i| > R_i(M) \tag{4.10}$$

for all $1 < i \leq n$. Supposing $\lambda \in \Gamma_{\mathbb{W}}(\mathscr{R}_1)_i$ for some fixed $1 < i \leq n$ and that $Q_i(M) \neq 0$, then (4.4) holds. Combining (4.4) with (4.9), we then have that

$$|\lambda_{11} L_1| \Big(|\lambda_{ii} L_i| - R_i(M) \Big) \leq 0.$$

Moreover, since $|\lambda_{ii} L_i| > R_i(M)$, from equation (4.10), this together with the previous inequality implies that $\lambda_{11} L_1$ must be zero. However, given that $\lambda_{11} L_1$ is a nonzero polynomial, this happens in at most finitely many values of $\lambda \in \mathbb{C}$. Similarly, the polynomial $Q_i(M)$ equals 0 on only a finite set of $\mathbb{C}$; hence the assumption that

$$\partial \Gamma_{\mathbb{W}}(M)_1 \setminus \bigcup_{j=2}^{n} \Gamma_{\mathbb{W}}(M)_j$$

is an infinite set in the complex plane yields a contradiction to assumption (4.8) for infinitely many points in this set. Hence, the result follows in the case that $S = \{2, \ldots, n\}$. By sequentially removing single indices from N, the result follows by repeated use of Theorem 1.3. $\square$

For a matrix $M \in \mathbb{W}_{\pi}^{n \times n}$, there is typically some region $\Gamma_{\mathbb{W}}(M)_i$ whose boundary is not contained in the union of the other j th Gershgorin regions. If this boundary is not a finite set of isolated points, then Theorem 4.4 guarantees that reducing M over $S = N - \{i\}$ strictly improves one's Gershgorin-type estimate of $\sigma(M)$.

This is demonstrated in the following example.

Example 4.4. Consider the matrix $M_0 \in \mathbb{W}_{\pi}^{5 \times 5}$ given by

$$M_0 = \begin{bmatrix} 0 & 0 & 1 & 0 & 1 \\ 0 & 0 & 0 & 1 & 1 \\ 0 & 1 & 0 & 0 & 0 \\ 1 & 0 & 0 & 0 & 0 \\ 1 & 1 & 1 & 1 & 0 \end{bmatrix}.$$

4.1 Gershgorin-Type Regions

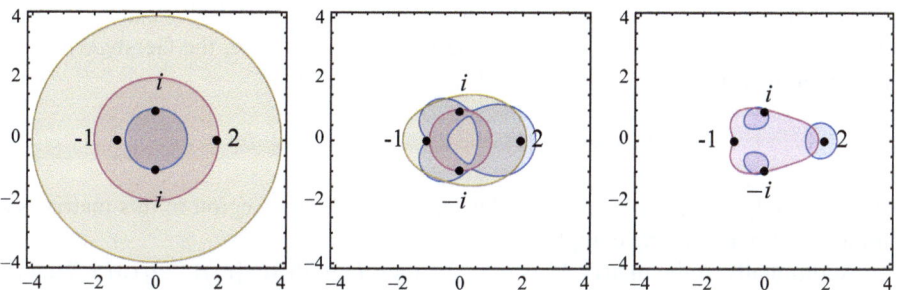

Fig. 4.4 Left: $\Gamma_W(M_0)$. Middle: $\Gamma_W(M_1)$. Right: $\Gamma_W(M_2)$, where in each, the spectrum $\sigma(M_0) = \{-1,-1,-i,i,2\}$ is indicated

Let $S_1 = \{1,2,3\}$ and $S_2 = \{1,2\}$. If $M_1 = \mathscr{R}(M_0; S_1)$ and $M_2 = \mathscr{R}(M_0; S_2)$, one can compute that

$$M_1 = \begin{bmatrix} \frac{\lambda+1}{\lambda^2} & \frac{1}{\lambda} & \frac{\lambda+1}{\lambda} \\ \frac{2\lambda+1}{\lambda^2} & \frac{1}{\lambda} & \frac{1}{\lambda} \\ 0 & 1 & 0 \end{bmatrix} \text{ and } M_2 = \begin{bmatrix} \frac{\lambda+1}{\lambda^2} & \frac{2\lambda+1}{\lambda^2} \\ \frac{2\lambda+1}{\lambda^2} & \frac{\lambda+1}{\lambda^2} \end{bmatrix}. \qquad (4.11)$$

The Gershgorin regions of M_0, M_1, and M_2 are shown in Fig. 4.4. Since

$$\partial \Gamma_W(M_0)_5 \setminus \bigcup_{j=1}^{4} \Gamma_W(M_0)_j \text{ and } \partial \Gamma_W(M_1)_3 \setminus \bigcup_{j=1}^{2} \Gamma_W(M_1)_j$$

consist of curves in $\mathbb{C}$, Theorem 4.4 implies the strict inclusions

$$\Gamma_W(M_2) \subset \Gamma_W(M_1) \subset \Gamma_W(M_0),$$

which can be seen in the figure. In addition, since

$$(M_0)_{\bar{S}_1 \bar{S}_1} = \begin{bmatrix} 0 & 0 \\ 0 & 1 \end{bmatrix} \text{ and } (M_1)_{\bar{S}_2 \bar{S}_2} = \begin{bmatrix} 0 & 0 & 0 \\ 0 & 0 & 0 \\ 1 & 1 & 0 \end{bmatrix},$$

it follows that $\sigma((M_0)_{\bar{S}_1 \bar{S}_1}) = \sigma((M_1)_{\bar{S}_2 \bar{S}_2}) = \{0\}$, not including multiplicities. Since the point 0 is in $\Gamma_W(M_1)$, $\Gamma_W(M_2)$, both $\Gamma_W(M_1)$ and $\Gamma_W(M_2)$ contain $\sigma(M_0)$. (Note that the matrix M_1 has been previously considered in Example 4.3.)

An important consequence of Theorem 4.3 is that a sequence of isospectral reductions on a matrix $M \in \mathbb{W}_\pi^{n \times n}$ can be used to obtain a sequence of estimates of $\sigma(M)$, each of which is better than the last. This can be seen, for instance, in Example 4.4. The extent to which a matrix M is reduced therefore determines the extent to which we have improved our estimate of $\sigma(M)$.

With this in mind, we note that the most that a matrix $M \in \mathbb{W}_\pi^{n\times n}$ can be reduced is over a single-element set $\{i\}$, for some $i \in N$. In this case, the Gershgorin-type region of the *fully reduced matrix* $\mathscr{R}(M;\{i\})$ is equal to

$$\Gamma_\mathbb{W}(\mathscr{R}(M;\{i\})) = \{\lambda \in \mathbb{C} : p(\lambda) = 0\}$$

for some polynomial $p(\lambda) \in \mathbb{C}[\lambda]$. Hence, the Gershgorin region of this matrix is a finite set of points in the complex plane.

Additionally, if the matrix M is in $\mathbb{C}^{n\times n}$, then $\Gamma_\mathbb{W}(\mathscr{R}(M;\{i\})) \subset \sigma(M)$. That is, the Gershgorin-type region of a fully reduced complex-valued matrix is a finite set of points that is contained in the matrix's spectrum.

As an example of a fully reduced matrix, i.e., a matrix reduced to a single index, let $M_3 = \mathscr{R}(M_2;\{1\})$, where M_2 is the 2×2 matrix considered in the previous example. For this matrix, we find that $\Gamma_\mathbb{W}(M_3) = \{-1,-1,-i,i,2\}$, which is, in fact, equal to the spectrum of the original unreduced matrix M_0 in this example.

Having considered how isospectral reductions can be used to improve eigenvalue estimates, we now turn our attention to how these techniques can be used to estimate the inverse spectrum of a matrix. Our main tool in this regard will be the spectral inverse operator introduced in Chap. 1.

If $M(\lambda) \in \mathbb{W}^{n\times n}$, then its inverse spectrum $\sigma^{-1}(M)$ comprises the complex numbers at which the determinant $\det(M - \lambda I)$ is undefined. Since the determinant of a matrix is composed of various products and sums of its entries, equations (1.1) and (1.2) imply the following proposition, in which $\overline{dom\,(M)}$ denotes the complement of $dom(M)$.

Proposition 4.1. *If $M(\lambda) \in \mathbb{W}^{n\times n}$, then $\sigma^{-1}(M) \subseteq \overline{dom\,(M)}$.*

Phrased another way, the inverse eigenvalues of a matrix $M \in \mathbb{W}^{n\times n}$ are complex numbers at which the matrix M is undefined, i.e., are contained in the complement of $dom\,(M)$. However, it is not always the case that an element of $\overline{dom\,(M)}$ is an inverse eigenvalue of M, as the following example shows.

Example 4.5. Consider the matrix $M \in \mathbb{W}^{2\times 2}$ given by

$$M(\lambda) = \begin{bmatrix} \frac{1}{\lambda-1} & \frac{1}{\lambda-1} \\ \frac{1}{\lambda} & \frac{\lambda+1}{\lambda} \end{bmatrix}.$$

As can be computed, $\sigma^{-1}(M) = \emptyset$, yet $\overline{dom\,(M)} = \{0,1\}$. That is, neither of the values $0, 1 \in \overline{dom\,(M)}$ is an inverse eigenvalue of M, although M is not defined at these points.

Proposition 4.1 gives us a way of estimating the inverse spectrum of a matrix $M \in \mathbb{W}^{n\times n}$. Alternatively, we can use Theorem 4.2 along with the spectral inverse $\mathscr{S}^{-1}(M)$ to generate a Gershgorin-type region that contains $\sigma^{-1}(M)$ if $M \in \mathbb{W}_\pi^{n\times n}$.

4.1 Gershgorin-Type Regions

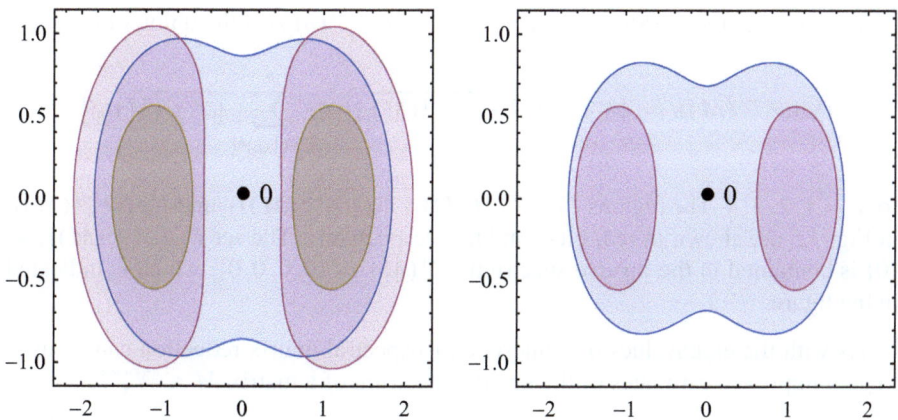

Fig. 4.5 *Left*: $\Gamma_{\mathbb{W}}(\mathscr{S}^{-1}(M))$. *Right*: $\Gamma_{\mathbb{W}}(R(\mathscr{S}^{-1}(M);\mathscr{B}))$ where the inverse spectrum $\sigma^{-1}(M) = \{0,0,0,0\}$ is indicated

Corollary 4.2. *Let $M(\lambda) \in \mathbb{W}_\pi^{n\times n}$. Then $\sigma^{-1}(M)$ is contained in the set*

$$\Gamma_{\mathbb{W}}(\mathscr{S}^{-1}(M)) = \bigcup_{i=1}^{n} \{\lambda \in \mathbb{C} : |\lambda - \overline{\mathscr{S}^{-1}(M)}_{ii}| \le r_i(\overline{\mathscr{S}^{-1}(M)})\}.$$

Example 4.6. Let $M \in \mathbb{W}_\pi^{4\times 4}$ be the matrix given by

$$M(\lambda) = \begin{bmatrix} \frac{1}{\lambda} & \frac{1}{\lambda} & 0 & 0 \\ 0 & \frac{1}{\lambda} & 1 & 0 \\ 0 & 0 & \frac{1}{\lambda} & 1 \\ 0 & 0 & 0 & \frac{1}{\lambda} \end{bmatrix}.$$

The polynomial extension of its spectral inverse is the matrix

$$\overline{\mathscr{S}^{-1}(M)} = \begin{bmatrix} -\lambda(\lambda^2-1)^9 & -\lambda(\lambda^2-1)^8 & -\lambda^2(\lambda^2-1)^7 & -\lambda^3(\lambda^2-1)^6 \\ 0 & -\lambda(\lambda^2-1)^5 & -\lambda^2(\lambda^2-1)^4 & -\lambda^3(\lambda^2-1)^3 \\ 0 & 0 & -\lambda(\lambda^2-1)^2 & -\lambda^2(\lambda^2-1) \\ 0 & 0 & 0 & -\lambda \end{bmatrix} + \lambda I,$$

which can be used to find the region $\Gamma_{\mathbb{W}}(\mathscr{S}^{-1}(M))$. This set is shown in Fig. 4.5 (left).

We note that the Gershgorin-type region $\Gamma_{\mathbb{W}}(\mathscr{S}^{-1}(M))$ is the union of the sets

$$\Gamma_{\mathbb{W}}(\mathscr{S}^{-1}(M))_i = \{\lambda \in \mathbb{C} : |\lambda - \overline{\mathscr{S}^{-1}(M)}_{ii}| \leq \sum_{j=1, j \neq i}^{n} |\overline{\mathscr{S}^{-1}(M)}_{ij}|\},$$

for $i = 1, 2, 3, 4$. The regions $\Gamma_{\mathbb{W}}(\mathscr{S}^{-1}(M))_1$, $\Gamma_{\mathbb{W}}(\mathscr{S}^{-1}(M))_2$, and $\Gamma_{\mathbb{W}}(\mathscr{S}^{-1}(M))_3$ in Fig. 4.5 are shown in red, blue, and tan, respectively. The set $\Gamma_{\mathbb{W}}(\mathscr{S}^{-1}(M))_4 = \{0\}$ is contained in the inverse spectrum $\sigma^{-1}(M) = \{0, 0, 0, 0\}$, which is indicated in the figure.

As with the eigenvalues of a matrix an isospectral matrix reduction can be used to gain improved estimates of the inverse spectrum of a matrix $M \in \mathbb{W}_\pi^{n \times n}$.

Theorem 4.5 (Improved Inverse Eigenvalue Estimates). *Let* $M(\lambda) \in \mathbb{W}_\pi^{n \times n}$, *where* $S \subset N$ *is nonempty. Then* $\Gamma_{\mathbb{W}}(\mathscr{R}(\mathscr{S}^{-1}(M); S)) \subseteq \Gamma_{\mathbb{W}}(\mathscr{S}^{-1}(M))$.

A proof of Theorem 4.5 can be obtained by following the proof of Theorem 4.3 using the fact that the spectral inverse $\mathscr{S}^{-1}(M)$ can be sequentially reduced to a unique matrix via Theorem 1.5.

Example 4.7. Let $M \in \mathbb{W}^{4 \times 4}$ be the matrix given in Example 4.6. For the index set $S = \{1, 2, 3\}$, the reduction of the spectral inverse of M is

$$\mathscr{R}(\mathscr{S}^{-1}(M); S) = \begin{bmatrix} \frac{-\lambda}{\lambda^2 - 1} & \frac{-\lambda}{(\lambda^2 - 1)^2} & \frac{-\lambda^2}{(\lambda^2 - 1)^3} \\ 0 & \frac{-\lambda}{\lambda^2 - 1} & \frac{-\lambda^2}{(\lambda^2 - 1)^2} \\ 0 & 0 & \frac{-\lambda}{\lambda^2 - 1} \end{bmatrix} + \lambda I.$$

Its polynomial extension is given by

$$\overline{\mathscr{R}(\mathscr{S}^{-1}(M); S)} = \begin{bmatrix} -\lambda(\lambda^2 - 1)^9 & -\lambda(\lambda^2 - 1)^8 & -\lambda^2(\lambda^2 - 1)^7 \\ 0 & -\lambda(\lambda^2 - 1)^5 & -\lambda^2(\lambda^2 - 1)^4 \\ 0 & 0 & -\lambda(\lambda^2 - 1)^2 \end{bmatrix} + \lambda I.$$

The Gershgorin-type region of the reduced matrix $\mathscr{R}(\mathscr{S}^{-1}(M); S)$ is shown in Fig. 4.5 (right), where one can see that $\sigma^{-1}(M) \subset \Gamma(\mathscr{R}(\mathscr{S}^{-1}(M); S)) \subset \Gamma(\mathscr{S}^{-1}(M))$. The regions $\Gamma(\mathscr{R}(\mathscr{S}^{-1}(M); S))_1$ and $\Gamma(\mathscr{R}(\mathscr{S}^{-1}(M); S))_2$ are shown in red and blue, respectively.

4.2 Brauer-Type Regions

By relating each row of a matrix $A \in \mathbb{C}^{n \times n}$ to a circle in the complex plane, Gershgorin's theorem allows for an algorithmically simple method for estimating the spectrum of a complex-valued matrix. One of the first successful attempts to improve on this result was given by Brauer.

4.2 Brauer-Type Regions

Similar to that of Gershgorin, Brauer's result associates the rows of a complex-valued matrix with a number of regions in the complex plane, the union of which contains the matrix's eigenvalues. The difference is that instead of each row being associated with a circle, as in Gershgorin's theorem, Brauer's regions are determined by pairs of rows and are oval in shape.

Theorem 4.6 (Brauer [27]). *Let $A \in \mathbb{C}^{n \times n}$, where $n \geq 2$. Then all eigenvalues of A are located in the set*

$$K(A) = \bigcup_{\substack{1 \leq i,j \leq n \\ i \neq j}} \{\lambda \in \mathbb{C} : |\lambda - A_{ii}||\lambda - A_{jj}| \leq r_i(A)r_j(A)\}. \tag{4.12}$$

Also, $K(A) \subseteq \Gamma(A)$.

The individual regions, given by $\{\lambda \in \mathbb{C} : |\lambda - A_{ii}||\lambda - A_{jj}| \leq r_i(A)r_j(A)\}$ in equation (4.12), are known as Cassini ovals and may consist of one or two distinct components. Moreover, there are $\binom{n}{2}$ such regions for every $n \times n$ matrix with complex entries. As with Gershgorin's theorem, we prove an extension to Brauer's theorem for matrices in $\mathbb{W}^{n \times n}$.

Theorem 4.7. *Let $M(\lambda) \in \mathbb{W}_\pi^{n \times n}$, where $n \geq 2$. Then $\sigma(M)$ is contained in the set*

$$K_\mathbb{W}(M) = \bigcup_{\substack{1 \leq i,j \leq n \\ i \neq j}} \{\lambda \in \mathbb{C} : |\lambda - \bar{M}_{ii}||\lambda - \bar{M}_{jj}| \leq r_i(\bar{M})r_j(\bar{M})\}.$$

Also, $K_\mathbb{W}(M) \subseteq \Gamma_\mathbb{W}(M)$.

Proof. As in the proof of Theorem 4.2, if $\alpha \in \sigma(M)$, then $\alpha \in \sigma(\bar{M})$, and the matrix $\bar{M}(\alpha)$ is in $\mathbb{C}^{n \times n}$. Brauer's theorem therefore implies that

$$|\alpha - \bar{M}(\alpha)_{ii}||\alpha - \bar{M}(\alpha)_{jj}| \leq r_i(\bar{M},\alpha)r_j(\bar{M},\alpha)$$

for some pair of distinct integers i and j. It then follows that $\alpha \in K_\mathbb{W}(M)$ or, $\sigma(M) \subseteq K_\mathbb{W}(M)$.

To prove the assertion that $K_\mathbb{W}(M) \subseteq \Gamma_\mathbb{W}(M)$, let

$$K_\mathbb{W}(M)_{ij} = \{\lambda \in \mathbb{C} : |\lambda - \bar{M}(\lambda)_{ii}||\lambda - \bar{M}(\lambda)_{jj}| \leq r_i(\bar{M},\lambda)r_j(\bar{M},\lambda)\} \tag{4.13}$$

for distinct i and j. The claim then is that $K_\mathbb{W}(M)_{ij} \subseteq \Gamma_\mathbb{W}(M)_i \cup \Gamma_\mathbb{W}(M)_j$. To see this, assume for a fixed $\lambda \in \mathbb{C}$ that $\lambda \in K_\mathbb{W}(M)_{ij}$ or,

$$|\lambda - \bar{M}(\lambda)_{ii}||\lambda - \bar{M}(\lambda)_{jj}| \leq r_i(\bar{M},\lambda)r_j(\bar{M},\lambda).$$

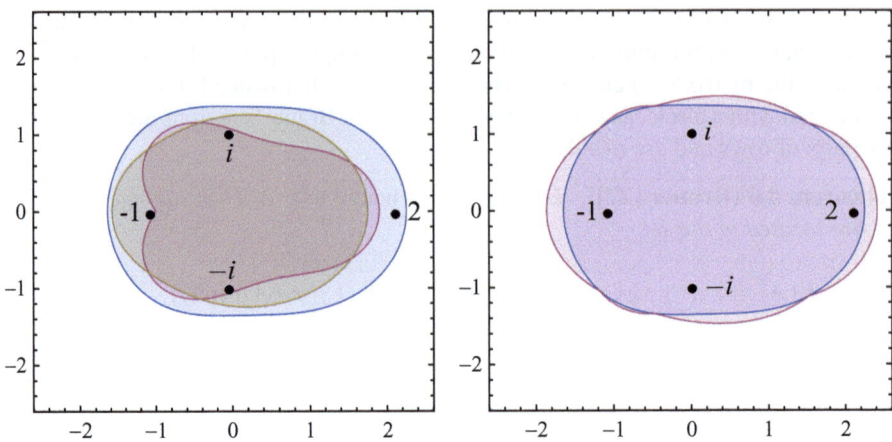

Fig. 4.6 *Left*: The Brauer region $K(M)$ for M from Example 4.3. *Right*: $K(M) \subseteq \Gamma(M)$

If $r_i(\bar{M}, \lambda) r_j(\bar{M}, \lambda) = 0$, then either $\lambda - \bar{M}(\lambda)_{ii} = 0$ or $\lambda - \bar{M}(\lambda)_{jj} = 0$. Since $\lambda = \bar{M}(\lambda)_{ii}$ implies $\lambda \in \Gamma_{\mathbb{W}}(M)_i$ and $\lambda = \bar{M}(\lambda)_{jj}$ implies $\lambda \in \Gamma_{\mathbb{W}}(M)_j$, we have $\lambda \in \Gamma_{\mathbb{W}}(M)_i \cup \Gamma_{\mathbb{W}}(M)_j$.

If $r_i(\bar{M}, \lambda) r_j(\bar{M}, \lambda) > 0$, then it follows that

$$\left(\frac{|\lambda - \bar{M}(\lambda)_{ii}|}{r_i(\bar{M}, \lambda)}\right)\left(\frac{|\lambda - \bar{M}(\lambda)_{jj}|}{r_j(\bar{M}, \lambda)}\right) \leq 1.$$

Since at least one of the two quotients on the left must be less than or equal to 1 we must have $\lambda \in \Gamma_{\mathbb{W}}(M)_i \cup \Gamma_{\mathbb{W}}(M)_j$, which verifies the claim, and the result follows. □

We call the region $K_{\mathbb{W}}(M)$ the *Brauer-type region* of the matrix M, and the region $K_{\mathbb{W}}(M)_{ij}$ given in (4.13), the *ijth Brauer-type region* of M. Applying Theorem 4.7 to the matrix M, given in Example 4.3, we have the Brauer-type region shown on the left-hand side of Fig. 4.6. On the right is a comparison between $K_{\mathbb{W}}(M)$ and $\Gamma_{\mathbb{W}}(M)$, where the inclusion $K_{\mathbb{W}}(M) \subseteq \Gamma_{\mathbb{W}}(M)$ is demonstrated. Here, $K_{\mathbb{W}(M)}$ is shown in blue, and $\Gamma_{\mathbb{W}}(M)$ in red.

As with Gershgorin's theorem, it is possible to improve a Brauer-type estimate of a matrix's eigenvalues by reducing the matrix.

Theorem 4.8 (Improved Brauer Regions). *Suppose $M(\lambda) \in \mathbb{W}_\pi^{n \times n}$. If $S \subset N$, where $|S| \geq 2$, then $K_{\mathbb{W}}(\mathscr{R}(M; S)) \subseteq K_{\mathbb{W}}(M)$.*

Theorem 4.8 has the following corollary.

Corollary 4.3. *If $M(\lambda) \in \mathbb{W}_\pi^{n \times n}$, where $S \subset N$ and $|S| \geq 2$, then*

$$K_{\mathbb{W}}(\mathscr{R}(M; S)) \subseteq K_{\mathbb{W}}(M) \cup \sigma(M_{\bar{S}\bar{S}}).$$

4.2 Brauer-Type Regions

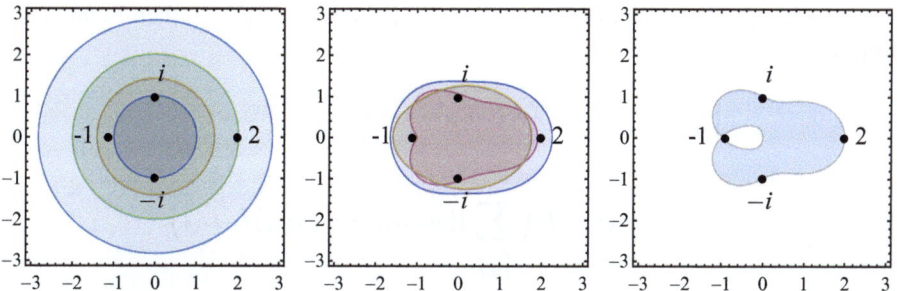

Fig. 4.7 *Left*: $K_W(M_0)$. *Middle*: $K_W(M_1)$. *Right*: $K_W(M_2)$, where in each the spectrum, $\sigma(M_0) = \{-1, -1, -i, i, 2\}$ is indicated

Proof. Since the region $K_W(\mathscr{R}(M;S))$ is a subset of $K_W(M)$, it follows from Theorem 1.1 that

$$\sigma(M) \subseteq K_W\big(\mathscr{R}(M;S)\big) \cup \sigma(M|_{\bar{S}\bar{S}}) \cup \sigma^{-1}(M).$$

Since the determinant $\det(M(\lambda) - \lambda I)$ is equal to $p(\lambda)/q(\lambda)$, where $p(\lambda), q(\lambda) \in \mathbb{C}[\lambda]$ have no common factors, it follows that $\sigma(M) \cap \sigma^{-1}(M) = \emptyset$, implying the result. $\square$

Continuing Example 4.4, the Brauer-type regions of M_0, M_1, and M_2 are shown in Fig. 4.7, where by Theorem 4.8, $K_W(M_2) \subseteq K_W(M_1) \subseteq K_W(M_0)$. Moreover, Theorem 4.7 implies $K_W(M_i) \subseteq \Gamma_W(M_i)$ for $i = 0, 1, 2$.

We note that if $M \in \mathbb{W}_\pi^{n \times n}$ is reduced over the index set S, where $|S| = m$, then there are $\binom{n}{2} - \binom{m}{2}$ fewer ijth Brauer-type regions to calculate in the reduced matrix $\mathscr{R}(M;S)$ than in M. Hence the number of regions quickly decreases as a matrix is reduced.

We now give a proof of Theorem 4.8.

Proof. Let $M \in \mathbb{W}_\pi^{n \times n}$ and $n \geq 3$. The claim is that

$$K_W(\mathscr{R}_1)_{ij} \subseteq K_W(M)_{1i} \cup K_W(M)_{1j} \cup K_W(M)_{ij} \tag{4.14}$$

for every pair $2 \leq i, j \leq n$, where $i \neq j$.

To see this, let $\lambda \in K_W(\mathscr{R}_1)_{ij}$ for fixed i and j, from which it follows that

$$\left|\left(\lambda_{ii} - \frac{\omega_{i1}\omega_{1i}}{\lambda_{11}}\right)L_i^1\right|\left|\left(\lambda_{jj} - \frac{\omega_{j1}\omega_{1j}}{\lambda_{11}}\right)L_j^1\right|$$

$$\leq \left(\sum_{\substack{\ell=2 \\ \ell \neq i}}^n \left|\left(\omega_{i\ell} + \frac{\omega_{i1}\omega_{1\ell}}{\lambda_{11}}\right)L_i^1\right|\right)\left(\sum_{\substack{\ell=2 \\ \ell \neq j}}^n \left|\left(\omega_{j\ell} + \frac{\omega_{j1}\omega_{1\ell}}{\lambda_{11}}\right)L_j^1\right|\right). \tag{4.15}$$

If we multiply both sides of (4.15) by $|\lambda_{11}q_{11}q_{i1}|$ and $|\lambda_{11}q_{11}q_{j1}|$, then Lemma 4.2 implies

$$\prod_{k=i,j} Q_k(M)|\lambda_{kk}\lambda_{11}L_1L_k - \omega_{k1}\omega_{1k}L_1L_k|$$

$$\leq \prod_{k=i,j} Q_k(M)\Big(\sum_{\substack{\ell=2 \\ \ell \neq k}}^n |(\omega_{k\ell}\lambda_{11} + \omega_{k1}\omega_{1\ell})L_1L_k|\Big).$$

Assuming for now that $Q_i(M)Q_j(M) \neq 0$ then by the triangle inequality

$$\prod_{k=i,j} \Big(|\lambda_{11}L_1\lambda_{kk}L_k| - |\omega_{1k}L_1\omega_{k1}L_k|\Big)$$
$$\leq \prod_{k=i,j}\Big(\sum_{\substack{\ell=2 \\ \ell \neq k}}^n |\lambda_{11}L_1\omega_{k\ell}L_k| + \sum_{\substack{\ell=2 \\ \ell \neq k}}^n |\omega_{1\ell}L_1\omega_{k1}L_k|\Big). \tag{4.16}$$

Suppose $\lambda \notin K_W(M)_{1i} \cup K_W(M)_{1j}$. Then $|\lambda_{11}L_1||\lambda_{kk}L_k| > R_1(M)R_k(M)$ for $k = i, j$. Moreover, if $|\lambda_{11}L_1| \leq R_1(M)$, then from (4.16), we obtain

$$\prod_{k=i,j}\Big(R_1(M)R_k(M) - |\omega_{1k}L_1\omega_{k1}L_k|\Big)$$
$$< \prod_{k=i,j}\Big(R_1(M)\sum_{\substack{\ell=2 \\ \ell \neq k}}^n |\omega_{k\ell}L_k| + \sum_{\substack{\ell=2 \\ \ell \neq k}}^n |\omega_{k1}L_1\omega_{1\ell}L_k|\Big). \tag{4.17}$$

Then, from the fact that

$$R_1(M)R_k(M) - |\omega_{k1}L_1\omega_{1k}L_k|$$
$$= R_1(M)\sum_{\substack{\ell=2 \\ \ell \neq k}}^n |\omega_{k\ell}L_k| + \sum_{\substack{\ell=2 \\ \ell \neq k}}^n |\omega_{k1}L_1\omega_{1\ell}L_k|, \tag{4.18}$$

it follows that (4.17) cannot hold. Therefore, if $\lambda \in K_W(\mathscr{R}_1)_{ij}$, $Q_i(M)Q_j(M) \neq 0$, and $\lambda \notin K_W(M)_{1i} \cup K_W(M)_{1j}$, then $|\lambda_{11}L_1| > R_1(M)$.

Proceeding as before, we assume that $\lambda \in K_W(\mathscr{R}_1)_{ij}$, so that in particular, (4.15) holds. Note that if $\lambda_{11} = 0$, then $\lambda \in K_W(M)_{1i} \cup K_W(M)_{1j}$, and the claim given in (4.14) holds. In what follows, we assume, then, that $\lambda_{11} \neq 0$. Moreover, if $Q_i(M)Q_j(M) \neq 0$, then multiplying both sides of (4.15) by $|\lambda_{11}q_{11}q_{i1}|$ and $|\lambda_{ii}L_iq_{11}q_{j1}|$ yields

4.2 Brauer-Type Regions

$$\left(|\lambda_{11}L_1\lambda_{ii}L_i| - |\omega_{1i}L_1\omega_{i1}L_i|\right)\left(|\lambda_{ii}L_i\lambda_{jj}L_jL_1| - |\omega_{1j}L_1\omega_{j1}L_j\frac{\lambda_{ii}L_i}{\lambda_{11}}|\right)$$

$$\leq \left(\sum_{\substack{\ell=2\\\ell\neq i}}^{n}|\lambda_{11}L_1\omega_{i\ell}L_i| + \sum_{\substack{\ell=2\\\ell\neq i}}^{n}|\omega_{1\ell}L_1\omega_{i1}L_i|\right)$$

$$\times \left(\sum_{\substack{\ell=2\\\ell\neq j}}^{n}|\lambda_{ii}L_i\omega_{j\ell}L_jL_1| + \sum_{\substack{\ell=2\\\ell\neq j}}^{n}|\omega_{1\ell}L_1\omega_{j1}L_j\frac{\lambda_{ii}L_i}{\lambda_{11}}|\right), \tag{4.19}$$

by use of the triangle inequality.

Supposing that $\lambda \notin K_W(M)_{1i} \cup K_W(M)_{ij}$, then we have both $R_1(M)R_i(M) < |\lambda_{11}L_1\lambda_{ii}L_i|$ and $R_i(M)R_j(M) < |\lambda_{ii}L_i\lambda_{jj}L_j|$. This together with (4.19) implies

$$\left(R_1(M)R_i(M) - |\omega_{1i}L_1\omega_{i1}L_i|\right)\left(R_i(M)R_j(M)L_1 - |\omega_{1j}L_1\omega_{j1}L_j\frac{\lambda_{ii}L_i}{\lambda_{11}}|\right)$$

$$< \left(\sum_{\substack{\ell=2\\\ell\neq i}}^{n}|\lambda_{11}L_1\omega_{i\ell}L_i| + \sum_{\substack{\ell=2\\\ell\neq i}}^{n}|\omega_{1\ell}L_1\omega_{i1}L_i|\right)$$

$$\times \left(|\lambda_{ii}L_iL_1|(R_j(M) - |\omega_{j1}L_j|) + |\omega_{j1}L_j\frac{\lambda_{ii}L_i}{\lambda_{11}}|(R_1(M) - |\omega_{1j}L_1|)\right).$$

If $|\lambda_{ii}L_i| \leq R_i(M)$, then

$$\left(R_1(M)R_i(M) - |\omega_{1i}L_1\omega_{i1}L_i|\right)\left(R_i(M)R_j(M)L_1 - |\omega_{1j}L_1\omega_{j1}L_j\frac{\lambda_{ii}L_i}{\lambda_{11}}|\right)$$

$$< \left(\sum_{\substack{\ell=2\\\ell\neq i}}^{n}|\lambda_{11}L_1\omega_{i\ell}L_i| + \sum_{\substack{\ell=2\\\ell\neq i}}^{n}|\omega_{1\ell}L_1\omega_{i1}L_i|\right)$$

$$\times \left(R_i(M)|L_1|(R_j(M) - |\omega_{j1}L_j|) + |\omega_{j1}L_j\frac{\lambda_{ii}L_i}{\lambda_{11}}|(R_1(M) - |\omega_{1j}L_1|)\right). \tag{4.20}$$

The claim then is that if $\lambda \notin K_W(M)_{1i} \cup K_W(M)_{1j}$, which implies $|\lambda_{11}L_1| > R_1(M)$ by the above, then the second term in each product of (4.20) has the relation

$$R_i(M)R_j(M) - |\omega_{1j}L_1\omega_{j1}L_j\frac{\lambda_{ii}L_i}{\lambda_{11}}|$$

$$\geq R_i(M)|L_1|(R_j(M) - |\omega_{j1}L_j|) + |\omega_{j1}L_j\frac{\lambda_{ii}L_i}{\lambda_{11}}|(R_1(M) - |\omega_{1j}L_1|). \tag{4.21}$$

110   4 Improved Eigenvalue Estimates

To see this, note that this is true if and only if

$$R_i(M)|\omega_{j1}L_jL_1| \geq |\omega_{j1}L_j\lambda_{ii}L_i|\frac{R_1(M)}{|\lambda_{11}|}.$$

Since this is true if and only if $|\lambda_{11}L_1|R_i(M) \geq R_1(M)|\lambda_{ii}L_i|$, this verifies that (4.21) holds, since we have both $R_i(M) \geq |\lambda_{ii}L_i|$ and $|\lambda_{11}L_1| > R_1(M)$. Therefore, equations (4.20) and (4.21) together imply that

$$R_1(M)R_i(M) - |\omega_{1i}L_1\omega_{i1}L_i| < \sum_{\substack{\ell=2 \\ \ell \neq i}}^n |\lambda_{11}L_1\omega_{i\ell}L_i| + \sum_{\substack{\ell=2 \\ \ell \neq i}}^n |\omega_{1\ell}L_1\omega_{i1}L_i|. \quad (4.22)$$

Rewriting the right-hand side of this inequality in terms of $R_k(M)$ (for $k = 1, i$) yields

$$R_1(M)R_i(M) < |\lambda_{11}L_1|R_i(M) - |\lambda_{11}L_1\omega_{i1}L_i| + |\omega_{i1}L_i|R_1(M).$$

This, in turn, implies that $R_i(M)(R_1(M) - |\lambda_{11}L_1|) < |\omega_{i1}L_i|(R_1(M) - |\lambda_{11}L_1|)$. However, it then follows that

$$R_i(M) = \sum_{\ell=1, \ell \neq i}^n |\omega_{i\ell}L_i| < |\omega_{i1}L_i|,$$

which is impossible.

Therefore, if both $Q_i(M)Q_j(M) \neq 0$ and $\lambda \notin K_W(M)_{1i} \cup K_W(M)_{1j} \cup K_W(M)_{ij}$, then $|\lambda_{ii}L_i| > R_i(M)$. Moreover, since this argument is symmetric in the indices i and j, it can be modified to show that if both

$$Q_i(M)Q_j(M) \neq 0 \text{ and } \lambda \notin K_W(M)_{1i} \cup K_W(M)_{1j} \cup K_W(M)_{ij},$$

then $|\lambda_{jj}L_j| > R_j(M)$.

With this in mind, by multiplying (4.15) by $|q_{11}q_{i1}|$ and $|q_{11}q_{i1}|$ and assuming once again that $Q_i(M)Q_j(M) \neq 0$, we see that the triangle inequality implies

$$\prod_{k=i,j} \left(|\lambda_{kk}L_k||L_1| - \left|\frac{\omega_{k1}\omega_{1k}}{\lambda_{11}}L_1L_k\right| \right)$$

$$\leq \prod_{k=i,j} \left(\sum_{\substack{\ell=1 \\ \ell \neq k}}^n |\omega_{k\ell}L_k||L_1| - |\omega_{k1}L_kL_1| + \sum_{\ell=2}^n \left|\frac{\omega_{k1}\omega_{1\ell}}{\lambda_{11}}L_kL_1\right| - \left|\frac{\omega_{k1}\omega_{1k}}{\lambda_{11}}L_kL_1\right| \right). $$
$$(4.23)$$

Hence, if $\lambda \notin K_W(M)_{1i} \cup K_W(M)_{1j} \cup K_W(M)_{ij}$, then from the previous calculations, $R_k(M) < |\lambda_{kk}L_k|$ for $k = 1, i, j$, which together with (4.23) implies that

$$\prod_{k=i,j}\left(R_k(M)|L_1| - |\frac{\omega_{k1}\omega_{1k}}{\lambda_{11}}L_1 L_k|\right)$$

$$< \prod_{k=i,j}\left(R_k(M)|L_1| - |\omega_{k1} L_k L_1| + |\omega_{k1} L_k|\frac{R_1(M)}{|\lambda_{11}|} - |\frac{\omega_{k1}\omega_{1k}}{\lambda_{11}}L_k L_1|\right).$$

Hence, for either $k = i$ or $k = j$, it follows that

$$-|\omega_{k1} L_k L_1| + |\omega_{k1} L_k|\frac{R_1(M)}{|\lambda_{11}|} > 0.$$

Therefore, $R_1(M) > |\lambda_{11} L_1|$, which is impossible. Since this implies that the value λ is not in $K_W(M)_{1i} \cup K_W(M)_{1j} \cup K_W(M)_{ij}$ unless $Q_i(M)Q_j(M) = 0$, suppose that this product is, in fact, equal to zero.

In that case, note that by modifying equation (4.5), we have

$$K_W(M)_{ij} = \left\{\lambda \in \mathbb{C} : \prod_{k=i,j}|L_{kk}(q_{kk}\lambda - p_{kk})| \leq \prod_{k=i,j}\left(\sum_{j=1,j\neq k}^n |p_{kj} L_{kj}|\right)\right\}$$

for $1 \leq k \leq n$. Hence if $Q_k(M) = 0$ for either $k = i$ or $k = j$, then by calculations analogous to those given in the proof of Theorem 4.3, it follows that $\lambda \in K_W(M)_{ik}$. This verifies the claim given in (4.14). Hence Theorem 4.8 holds for $S = N - \{1\}$.

As in the previous proofs, Theorem 1.3 can be invoked to generalize this result to the reduction over the any index set S, as long as $|S| \geq 2$. □

The following is the reason Brauer-type estimates improve under reduction. If we reduce $M \in \mathbb{W}_\pi^{n\times n}$ over the set $S = N - \{k\}$, then the ijth Brauer region of the reduced matrix is contained in the ijth, ikth, and jkth Brauer regions of the unreduced matrix. As an example of this, the inclusion

$$K_W(\mathscr{R}(M;S))_{23} \subseteq K_W(M)_{12} \cup K_W(M)_{13} \cup K_W(M)_{23}$$

is shown in Fig. 4.3 (right) for the matrix M given in (4.7) and index set $S = \{2, 3\}$.

4.3 Brualdi-Type Regions

The eigenvalue estimates of both Gershgorin and Brauer use a technique of associating regions in the complex plane with the rows and pairs of rows of a matrix $M \in \mathbb{C}^{n\times n}$. Brualdi [5] was able to improve on these results using combinations of the matrix rows corresponding to the cycle structure of the graph G associated with the matrix M. This result was later extended by Varga [27], who likewise used the cycles of G to estimate the spectrum of M.

In this section, we show that the results of Brualdi and Varga can be extended to matrices in $\mathbb{W}^{n \times n}$. Because of the graph-theoretic nature of these results, we take the point of view in this section that we are considering the spectrum of a graph $G \in \mathbb{G}$ instead of a matrix $M \in \mathbb{W}^{n \times n}$. This shift in perspective from Sections 4.1 and 4.2 will require some changes in our notation. However, we will endeavor to keep this as uniform as possible.

For notational convenience, we let $\mathbb{G}^n$ be the set of graphs

$$\mathbb{G}^n = \{G = (V, E, \omega) : M(G) \in \mathbb{W}^{n \times n}\},$$

or the graphs with n vertices for some $n \in \mathbb{N}$. Hence $\mathbb{G} = \cup_{n \geq 1} \mathbb{G}^n$. We note that there is a one-to-one correspondence between the graphs in $\mathbb{G}^n$ and the matrices $\mathbb{W}^{n \times n}$. Therefore, we may talk of a graph $G \in \mathbb{G}^n$ associated with a matrix $M = M(G)$ in $\mathbb{W}^{n \times n}$ and conversely without ambiguity.

Let $G = (V, E, \omega)$ be a graph in $\mathbb{G}^n$. A *strong cycle* of G is a cycle $v_1, \ldots, v_m$ such that $m \geq 2$. Furthermore, if $v_i \in V$ has no strong cycle passing through it, then we define its associated *weak cycle* as v_i, regardless of whether $e_{ii} \in E$. For $G \in \mathbb{G}$, we let $C_s(G)$ and $C_w(G)$ denote the sets of strong and weak cycles of G, respectively, and let $C(G) = C_s(G) \cup C_w(G)$ denote the *cycle set* of G.

Recall from Chap. 3 that a directed graph G is *strongly connected* if there is a path from each vertex of the graph to every other vertex. Moreover, the *strongly connected components* of $G = (V, E, \omega)$ are its maximal strongly connected subgraphs. Moreover, the vertex set $V = \{v_1, \ldots, v_n\}$ of G can always be labeled in such a way that $M(G)$ has the triangular block structure

$$M(G) = \begin{bmatrix} M(\mathbb{S}_1(G)) & 0 & \cdots & 0 \\ * & M(\mathbb{S}_2(G)) & & \vdots \\ \vdots & & \ddots & 0 \\ * & & \cdots & * \; M(\mathbb{S}_m(G)) \end{bmatrix},$$

where $\mathbb{S}_i(G)$ is a strongly connected component of G, and $*$ are block matrices with possibly nonzero entries (see [6, 22, 27] for more details).

Since the strongly connected components of a graph are unique, then for $G \in \mathbb{G}^n$, we define

$$\tilde{r}_i(G) = \sum_{j \in N_\ell, j \neq i} |M(\mathbb{S}_\ell(G))_{ij}| \text{ for } 1 \leq i \leq n,$$

where $i \in N_\ell$, and N_ℓ is the set of indices indexing the vertices in $\mathbb{S}_\ell(G)$. That is, $\tilde{r}_i(G)$ is the ith absolute row sum of $M(G)$ restricted to the strongly connected

4.3 Brualdi-Type Regions

component containing v_i. We let $\bar{G}$ be the graph with adjacency matrix $\overline{M(G)}$. Furthermore, we let $\tilde{r}_i(\bar{G}) = \tilde{r}_i(\bar{G}, \lambda)$, where we consider $\tilde{r}_i(\bar{G}, \cdot) : \mathbb{C} \to \mathbb{C}$.

If $A \in \mathbb{C}^{n \times n}$, then we write $\tilde{r}_i(G, \lambda) = \tilde{r}_i(A)$, where $A = M(G)$. Moreover, we let $C(A) = C_s(A) \cup C_w(A)$. This allows us to state the following theorem of Varga.

Theorem 4.9 (Varga [27]). *Let $A \in \mathbb{C}^{n \times n}$. Then the eigenvalues of A are contained in the set*

$$B(A) = \bigcup_{\gamma \in C(A)} \{\lambda \in \mathbb{C} : \prod_{v_i \in \gamma} |\lambda - A_{ii}| \leq \prod_{v_i \in \gamma} \tilde{r}_i(A)\}.$$

Also, $B(A) \subseteq K(A)$.

As with the theorems of Gershgorin and Brauer, this result can be extended to matrices in $\mathbb{W}^{n \times n}$.

Theorem 4.10. *Let $G \in \mathbb{G}$. Then $\sigma(G)$ is contained in the set*

$$B_{\mathbb{W}}(G) = \bigcup_{\gamma \in C(\bar{G})} \{\lambda \in \mathbb{C} : \prod_{v_i \in \gamma} |\lambda - M(\bar{G})_{ii}| \leq \prod_{v_i \in \gamma} \tilde{r}_i(\bar{G})\}. \quad (4.24)$$

Also, $B_{\mathbb{W}}(G) \subseteq K_{\mathbb{W}}(G)$.

We call $B_{\mathbb{W}}(G)$ the *Brualdi-type region* of the graph G, and the set

$$B_{\mathbb{W}}(G)_\gamma = \{\lambda \in \mathbb{C} : \prod_{v_i \in \gamma} |\lambda - M(\bar{G})_{ii}| \leq \prod_{v_i \in \gamma} \tilde{r}_i(\bar{G})\},$$

the Brualdi-type region associated with the cycle $\gamma \in C(\bar{G})$.

Proof. For $G \in \mathbb{G}^n$, let $\bar{G} = \bar{G}(\lambda)$, where for fixed $\alpha \in \mathbb{C}$, $\bar{G}(\alpha)$ is the graph with adjacency matrix $M(\bar{G}, \alpha) \in \mathbb{C}^{n \times n}$. Moreover, for every $\gamma = \{v_1, \ldots, v_m\}$ in $C(\bar{G})$ and fixed $\alpha \in \mathbb{C}$, let $\gamma(\alpha)$ be the set of vertices $\{v_1, \ldots, v_m\}$ in the graph $\bar{G}(\alpha)$.

Using this notation, if $\alpha \in \sigma(G)$, then Lemma 4.1 and Theorem 4.9 imply that there exists $\gamma' \in C(\bar{G}(\alpha))$ such that

$$\prod_{v_i \in \gamma'} |\alpha - M(\bar{G}, \alpha)_{ii}| \leq \prod_{v_i \in \gamma'} \tilde{r}_i(\bar{G}, \alpha). \quad (4.25)$$

There are then two possibilities: either $\gamma' \in C(\bar{G})$ or $\gamma' \notin C(\bar{G})$. If $\gamma' \in C(\bar{G})$, then the set of vertices $\gamma'(\alpha)$ is also a cycle in $\bar{G}$, in which case, relations (4.24) and (4.25) imply $\alpha \in B_{\mathbb{W}}(G)$. Suppose, then, that $\gamma' \notin C(\bar{G})$.

Note that if $\gamma' \in C_s(\bar{G}(\alpha))$, then since $M(\bar{G}, \alpha)_{ij} \neq 0$ implies $M(\bar{G}, \lambda)_{ij} \neq 0$ for $i \neq j$, it follows that $\gamma' \in C_s(\bar{G})$. Since this is impossible, $\gamma' \in C_w(\bar{G}(\alpha))$ or γ' must be a loop of some vertex v_j, where the graph induced by $\{v_j\}$ in $\bar{G}(\alpha)$

is a strongly connected component of $\bar{G}(\alpha)$. Thus, equation (4.25) is equivalent to $|\alpha - M(\bar{G}, \alpha)_{jj}| \leq 0$, implying $\alpha = M(\bar{G}, \alpha)_{jj}$.

Since some cycle $\gamma \in C(\bar{G})$ contains the vertex v_j, it follows that α is contained in the set

$$\{\lambda \in \mathbb{C} : \prod_{v_i \in \gamma} |\lambda - M(\bar{G}, \lambda)_{ii}| \leq \prod_{v_i \in \gamma} \tilde{r}_i(\bar{G}, \lambda)\},$$

implying that $\alpha \in B_{\mathbb{W}}(G)$.

To show now that $B_{\mathbb{W}}(G) \subseteq K_{\mathbb{W}}(G)$, let $\gamma \in C(\bar{G})$. Supposing $\gamma \in C_w(\bar{G})$, then $\gamma = \{v_i\}$ for some vertex v_i of G and

$$B_{\mathbb{W}}(G)_\gamma = \{\lambda \in \mathbb{C} : |\lambda - M(\bar{G}, \lambda)_{ii}| = 0\},$$

since v_i is the vertex set of some strongly connected component of $\bar{G}$. It follows from (4.13) that $B_{\mathbb{W}}(G)_\gamma \subseteq K_{\mathbb{W}}(G)_{ij}$ for every $1 \leq j \leq n$, where $i \neq j$. In particular, note that if $\tilde{r}_i(\bar{G}, \lambda) = 0$, then $\lambda \in K_{\mathbb{W}}(G)_{ij}$ for every $1 \leq j \leq n$, where $i \neq j$.

If, on the other hand, $\gamma \in C_s(\bar{G})$, then for convenience, let $\gamma = \{v_1, \ldots, v_p\}$, where $p > 1$, and note that

$$B_{\mathbb{W}}(G)_\gamma = \{\lambda \in \mathbb{C} : \prod_{i=1}^{p} |\lambda - M(\bar{G}, \lambda)_{ii}| \leq \prod_{i=1}^{p} \tilde{r}_i(\bar{G}, \lambda)\}. \quad (4.26)$$

Assuming $0 < \tilde{r}_i(\bar{G}, \lambda)$ for all $1 \leq i \leq p$, then for fixed $\lambda \in B_{\mathbb{W}}(G)_\gamma$, it follows by raising both sides of the inequality in (4.26) to the $(p-1)$st power that

$$\prod_{\substack{1 \leq i,j \leq p \\ i \neq j}} \left(\frac{|\lambda - M(\bar{G}, \lambda)_{ii}||\lambda - M(\bar{G}, \lambda)_{jj}|}{\tilde{r}_i(\bar{G}, \lambda)\tilde{r}_j(\bar{G}, \lambda)} \right) \leq 1. \quad (4.27)$$

Since not all the terms of the product in (4.27) can exceed unity, it follows that for some pair of indices ℓ and k, where $1 \leq \ell, k \leq p$ and $\ell \neq k$, we must have

$$|\lambda - M(\bar{G}, \lambda)_{kk}||\lambda - M(\bar{G}, \lambda)_{\ell\ell}| \leq \tilde{r}_k(\bar{G}, \lambda)\tilde{r}_\ell(\bar{G}, \lambda). \quad (4.28)$$

Using the fact that $\tilde{r}_i(\bar{G}, \lambda) \leq r_i(\bar{G}, \lambda)$ for all $1 \leq i \leq n$, we conclude that $\lambda \in K_{\mathbb{W}}(G)_{k\ell}$, completing the proof. □

The Brualdi-type region for the graph G with adjacency matrix $M = M(G)$, given by (4.2), is shown in Fig. 4.8. We note that $B_{\mathbb{W}}(G) = K_{\mathbb{W}}(M)$ in this particular case.

We now consider Brualdi's original result, which can be stated as follows.

4.3 Brualdi-Type Regions

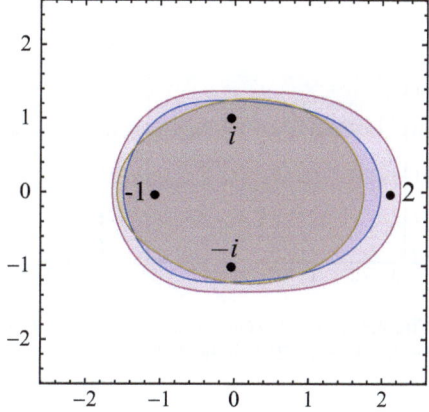

Fig. 4.8 The Brualdi-type region $B_{\mathbb{W}}(G)$ for G in Fig. 4.2 (*left*)

Theorem 4.11 (Brualdi [5]). *Let $A \in \mathbb{C}^{n \times n}$, where $C_w(A) = \emptyset$. Then the eigenvalues of A are contained in the set*

$$br(A) = \bigcup_{\gamma \in C(A)} \{\lambda \in \mathbb{C} : \prod_{v_i \in \gamma} |\lambda - A_{ii}| \leq \prod_{v_i \in \gamma} r_i(A)\}.$$

Also, $B_{\mathbb{W}}(A) \subseteq br_{\mathbb{W}}(A) \subseteq K_{\mathbb{W}}(A)$.

As with the theorems of Gershgorin, Brauer, and Varga, this result generalizes to matrices with entries in $\mathbb{W}$ as follows.

Theorem 4.12. *Let $G \in \mathbb{G}$, where $C_w(G) = \emptyset$. Then $\sigma(G)$ is contained in the set*

$$br_{\mathbb{W}}(G) = \bigcup_{\gamma \in C(\bar{G})} \{\lambda \in \mathbb{C} : \prod_{v_i \in \gamma} |\lambda - M(\bar{G})_{ii}| \leq \prod_{v_i \in \gamma} r_i(\bar{G})\}. \quad (4.29)$$

Also, $B_{\mathbb{W}}(G) \subseteq br_{\mathbb{W}}(G) \subseteq K_{\mathbb{W}}(G)$.

Proof. Note that for every graph $G \in \mathbb{G}$, we have $\tilde{r}_i(\bar{G}) \leq r_i(\bar{G})$ for all $\lambda \in \mathbb{C}$. Hence

$$B_{\mathbb{W}}(G) \subseteq \bigcup_{\gamma \in C(\bar{G})} \{\lambda \in \mathbb{C} : \prod_{v_i \in \gamma} |\lambda - M(\bar{G})_{ii}| \leq \prod_{v_i \in \gamma} r_i(\bar{G})\}.$$

Theorem 4.10 then implies that $\sigma(G)$ is contained in the set $br_{\mathbb{W}}(G)$. Furthermore, if $\tilde{r}_i(G)$ is replaced by $r_i(G)$ in the proof of Theorem 4.10, then (4.28) implies that $br_{\mathbb{W}}(G) \subseteq K_{\mathbb{W}}(G)$, completing the proof. □

We will refer to the region $br_{\mathbb{W}}(G)$, given in (4.29), as the *original Brualdi-type region* of G.

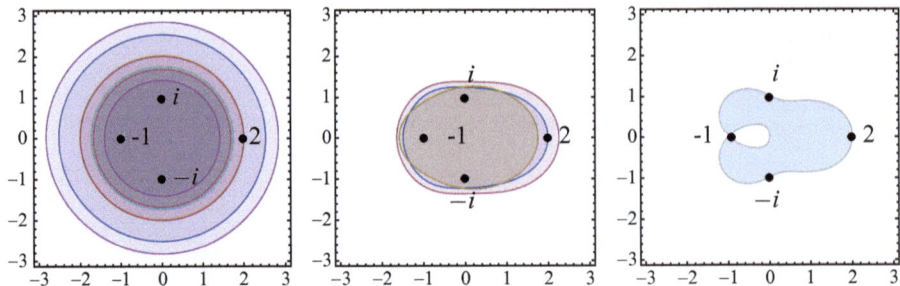

Fig. 4.9 *Left*: $B_{\mathbb{W}}(G_0)$. *Middle*: $B_{\mathbb{W}}(G_1)$. *Right*: $B_{\mathbb{W}}(G_2)$, where in each, the spectrum $\sigma(G_0) = \{-1, -1, -i, i, 2\}$ is indicated

As demonstrated in sections 4.1 and 4.2, the eigenvalue regions associated with both Gershgorin and Brauer shrink under isospectral reduction. However, for both Brualdi-type and original Brualdi-type regions, the situation is more complicated. For certain reductions, the Brualdi-type (original Brualdi-type) region of a graph may decrease in size similarly to Gershgorin-type and Brauer-type regions. In other cases, the Brualdi-type (original Brualdi-type) region of a graph may increase in size as the graph is reduced.

Example 4.8. For example, suppose G_i is the graph with adjacency matrix $M(G_i) = M_i$ considered in Example 4.4 for $i = 0, 1, 2$. Here, the graph G_{i+1} is the reduction of the graph G_i for $i = 0, 1$. In this particular case, we have the inclusions

$$B_{\mathbb{W}}(G_2) \subseteq B_{\mathbb{W}}(G_1) \subseteq B_{\mathbb{W}}(G_0),$$

which can be seen in Fig. 4.9. That is, the Brualdi-type regions of the graph G_0 shrink under this sequence of isospectral reductions.

In the following, we give an example of a graph whose Brualdi-type regions do not have this property.

Example 4.9. Consider the graph $H \in \mathbb{G}_\pi$ given in Fig. 4.10. If H is reduced over the set $S = \bar{v}_1 = \{v_2, v_3, v_4\}$ and $T = \bar{v}_4 = \{v_1, v_2, v_3\}$, then

$$M(\mathscr{R}_S(H)) = \begin{bmatrix} \frac{1}{\lambda} & \frac{1}{10} & 0 \\ \frac{10}{\lambda} & 0 & 1 \\ 0 & 1 & 0 \end{bmatrix} \text{ and } M(\mathscr{R}_T(H)) = \begin{bmatrix} \frac{1}{\lambda} & \frac{1}{\lambda} & 0 \\ 1 & 0 & 1 \\ 0 & 1 & 0 \end{bmatrix}.$$

In this case, we have the strict inclusions

$$B_{\mathbb{W}}(\mathscr{R}_T(H)) \subset B_{\mathbb{W}}(H) \subset B_{\mathbb{W}}(\mathscr{R}_S(H)),$$

4.3 Brualdi-Type Regions

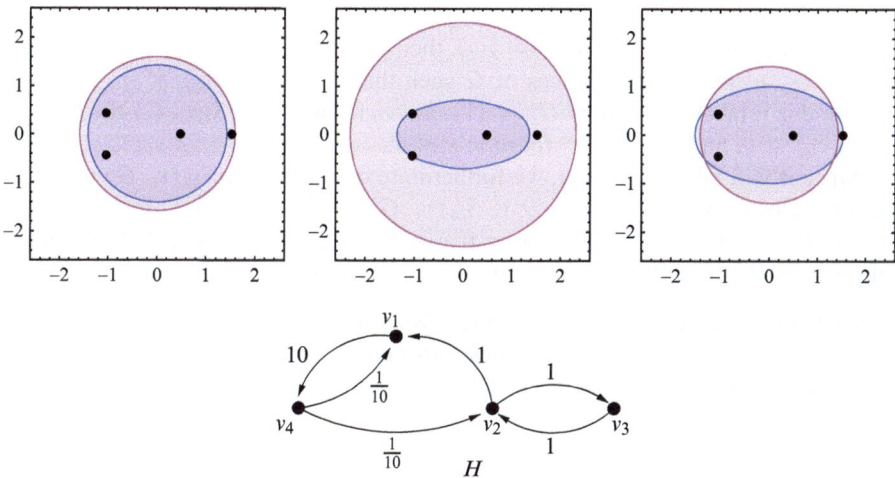

Fig. 4.10 Top Left: $B_{\mathbb{W}}(H)$. Top Middle: $B_{\mathbb{W}}(\mathscr{R}_S(H))$. Top Right: $B_{\mathbb{W}}(\mathscr{R}_T(H))$, where $S = \bar{v}_1$ and $T = \bar{v}_4$. The eigenvalues $\sigma(H)$ are indicated

shown in Fig. 4.10. In particular, since $B_{\mathbb{W}}(H) \subset B_{\mathbb{W}}(\mathscr{R}_S(H))$, then reducing the graph H over S increases the size of the graph's Brualdi-type region. Therefore, isospectral graph reductions do not always improve Brualdi-type estimates.

The fact that isospectral graph reductions do not always improve Brualdi-type regions should not be too surprising. The major reason, as we will see, is that isospectral graph reductions do not preserve the cycle structure of a graph and may both create and destroy cycles. Since Brualdi's eigenvalue estimates rely on this structure, this is a very different technique for finding the eigenvalues of a matrix from the methods of Gershgorin and Brauer, which do not take into account any particular structure.

To give a sufficient condition under which a graph's Brualdi-type region shrinks as the graph is reduced, we require the following definitions. First, let $G = (V, E, \omega)$, where $G \in \mathbb{G}_\pi^n$ for some $n \geq 1$, and let G have the strongly connected components $\mathbb{S}_1(G), \ldots, \mathbb{S}_m(G)$. Define

$$E^{scc} = \{e \in E : e \in \mathbb{S}_i(G), 1 \leq i \leq m\}.$$

The cycle $\gamma \in C(G)$ is said to be *adjacent* to $v_i \in V$ if $v_i \notin \gamma$ and there is some vertex $v_j \in \gamma$ such that $e_{ji} \in E^{scc}$.

Second, for every $v_i \in V$, we define the set of cycles

$$\text{Adj}(v_i, G) = \{\gamma \in C(G) : \gamma \text{ is adjacent to } v_i\}.$$

Moreover, if $C(v_i, G) = \{\gamma \in C(G) : v_i \in \gamma\}$, then let $S(v_i, G) \subseteq C(v_i, G)$ be the set defined as follows.

For a fixed $i \in N$, let $\gamma = v_{\alpha_1}, \ldots, v_{\alpha_m}$ be a cycle in $C(v_i, G)$, where $n \geq m \geq 1$ and $v_i = v_{\alpha_1}$. If $m = 1$, that is, $\gamma = \{v_i\}$, then $\gamma \in S(v_i, G)$. Otherwise, supposing $1 < m \leq n$, relabel the vertices of G such that v_{α_j} is v_j for $1 \leq j \leq m$ and denote this relabeled graph by $G_r = (V_r, E_r, \omega_r)$. Then $\gamma \in S(v_1, G)$ if $e_{j1} \notin E_r$ for $1 < j < m$ and $e_{mk} \notin E_r^{scc}$ for $m < k \leq n$.

Since it will be needed later, we furthermore define the set $S_{br}(v_i, G)$ to be the set of cycles in $S(v_i, G)$, where $\gamma \in S_{br}(v_1, G)$ if $e_{j1} \notin E_r$ for $1 < j < m$ and $e_{mk} \notin E_r$ for $m < k \leq n$. Moreover, if $v \in V$, we let $\bar{v} = V - \{v\}$. With this in place, we state the following theorem.

Theorem 4.13 (Improved Brualdi-Type Regions). *Let $G = (V, E, \omega)$, where $G \in \mathbb{G}_\pi$ and $|V| \geq 2$. If $v \in V$ such that both $Adj(v, G) = \emptyset$ and $C(v, G) = S(v, G)$, then $B_{\mathbb{W}}(\mathscr{R}_{\bar{v}}(G)) \subseteq B_{\mathbb{W}}(G)$.*

Theorem 4.13 states that if the vertex v is adjacent to no cycle in $C(G)$, and each cycle passing through v is in $S(v, G)$, then removing this vertex from the graph will improve its Brualdi-type region.

We note that for the graph H in Fig. 4.10, the vertex v_1 has the property that $Adj(v_1, H) = \{v_2, v_3\} \neq \emptyset$. Hence, Theorem 4.13 does not apply to the reduction of H over $S = \bar{v}_1$. On the other hand, the vertex v_4 of H has the property that $Adj(v_4, H) = \emptyset$ and $S(v_4, H) = C(v_4, H)$. Therefore, reducing H over the vertex set $T = \bar{v}_4$ improves the Brualdi-type region of this graph, which can be seen by comparing the upper left-hand and upper right-hand sides of Fig. 4.10.

Example 4.10. To see why the condition $C(v, G) = S(v, G)$ is necessary in Theorem 4.13, we consider the following. Let $J, \mathscr{R}_S(J) \in \mathbb{G}$ be the matrices given by

$$M(J) = \begin{bmatrix} 0 & 1 & 0 & 0 \\ 0 & 0 & 1 & 0 \\ 1 & 0 & 0 & 1 \\ 1 & 0 & 0 & 0 \end{bmatrix} \text{ and } M(\mathscr{R}_S(J)) = \begin{bmatrix} 0 & 1 & 0 \\ \frac{1}{\lambda} & 0 & 1 \\ \frac{1}{\lambda} & 0 & 0 \end{bmatrix},$$

where $S = \bar{v}_1 = \{v_2, v_3, v_4\}$. In this case, the region $B_{\mathbb{W}}(\mathscr{R}_S(J))$ is not a subset of $B_{\mathbb{W}}(J)$. We note that $Adj(v_1, J) = \emptyset$, but $S(v_1, J)$ consists of the cycle v_1, v_2, v_3, whereas the cycle set $C(v_1, J)$ is equal to $\{v_1, v_2, v_3; v_1, v_2, v_3, v_4\}$. That is, $C(v_1, J) \neq S(v_1, J)$.

Observe that graph reductions can increase, decrease, or maintain the number of cycles in a graph's cycle set. For instance, the graph G_0 in Example 4.8 has 12 cycles in its cycle set, whereas G_1 has 3, and G_2 has 1 (see Fig. 4.9). In contrast, let $P, \mathscr{R}_U(P) \in \mathbb{G}$ be the graphs with adjacency matrices

4.3 Bruladi-Type Regions

$$M(P) = \begin{bmatrix} 0 & 1 & 0 & 0 & 0 \\ 0 & 0 & 0 & 0 & 1 \\ 0 & 0 & 0 & 1 & 0 \\ 0 & 0 & 0 & 0 & 1 \\ 1 & 0 & 1 & 0 & 0 \end{bmatrix} \text{ and } M(\mathscr{R}_U(P)) = \begin{bmatrix} 0 & 1 & 0 & 0 \\ \frac{1}{\lambda} & 0 & \frac{1}{\lambda} & 0 \\ 0 & 0 & 0 & 1 \\ \frac{1}{\lambda} & 0 & \frac{1}{\lambda} & 0 \end{bmatrix},$$

where $U = \{v_1, v_2, v_3, v_4\}$. Here, the cycle set $C(P)$ is equal to $\{v_1, v_2, v_5; v_3, v_3, v_5\}$, whereas $C(\mathscr{R}_U(P)) = \{v_1, v_2; v_3, v_4; v_1, v_2, v_3, v_4\}$. That is, reducing P over U increases the number of cycles needed to compute the associated Brualdi-type region from 2 to 3. Hence, in this case the number of subregions that makeup the Brualdi-type region increase as the graph is reduced. This is in contrast to Gershgorin and Brauer-type regions in which the number of subregions always decrease as the associated graph is reduced.

Before continuing on to Brualdi's original result, we note that as with matrices, it is possible to *fully reduce* a graph $G \in \mathbb{G}_\pi$ to a single vertex. That is, if $S = \{v_i\}$, then the graph $\mathscr{R}_S(G)$ consists of a single vertex v_i and has the Brualdi-type region

$$\mathbb{B}_W(\mathscr{R}_S(G)) = \{\lambda \in \mathbb{C} : p(\lambda) = 0\}$$

for some polynomial $p(\lambda) \in \mathbb{C}[\lambda]$. Using equation (1.11), one can show that

$$\mathbb{B}_W(\mathscr{R}_S(G)) \subseteq \sigma(G) \cup \sigma^{-1}(G|_{\bar{S}}),$$

and so in the case of a fully reduced graph, the associated Brualdi-type region is a finite collection of points in the complex plane.

Moreover, if the graph G has complex-valued weights, then $\sigma^{-1}(G) = \emptyset$. This together with Theorem 4.10 implies that $\mathbb{B}_W(\mathscr{R}_S(G)) \subseteq \mathbb{B}_W(G)$. This fact leads to the following remark.

Remark 4.1. If $G = (V, E, \omega)$ is a graph with complex-valued weights, then there is always some $S \subset V$ such that $\mathbb{B}_W(\mathscr{R}_S(G)) \subseteq \mathbb{B}_W(G)$. That is, it is always possible to improve the Brualdi-type region of G by reducing G over some subset of its vertices.

In the case of Brualdi's original result (Theorem 4.12), we must deal with the following complications. First, for a given graph $G \in \mathbb{G}_\pi$, where $C_w(G) = \emptyset$, it may not be the case that $C_w(\mathscr{R}_{\bar{v}}(G)) = \emptyset$. Furthermore, since the edges between strongly connected components play a role in the associated eigenvalue inclusion region (see (4.29)), this also complicates whether estimates given by the original Brualdi-type region improve as the graph is reduced. However, it is possible to give sufficient conditions under which this is the case.

Theorem 4.14 (Improved Original Brualdi-Type Regions). *Let $G = (V, E, \omega)$ be in $\mathbb{G}_\pi$ and $v \in V$. If $Adj(v, G) = \emptyset$, $C(v, G) = S_{br}(v, G)$, and both of the sets $C_w(G)$ and $C_w(\mathscr{R}_{\bar{v}}(G))$ are empty, then $br_W(\mathscr{R}_{\bar{v}}(G)) \subseteq br_W(G)$.*

We finish this section now by giving a proof of Theorem 4.13, and then a proof of Theorem 4.14. In order to prove Theorem 4.13, we first give the following lemma.

Lemma 4.3. *Let $G \in \mathbb{G}_\pi^n$ for $n \geq 2$ and suppose that $\mathrm{Adj}(v_1, G) = \emptyset$ and $C(v_1, G) = S(v_1, G)$. Moreover, let $\gamma = \{v_1, \ldots, v_m\}$ and $\gamma' = \{v_2, \ldots, v_m\}$ for $m \geq 2$. If $\gamma \in C(G)$ and $\gamma' =\in C(\mathscr{R}_1(G))$, then $B_\mathbb{W}(\mathscr{R}_1(G))_{\gamma'} \subseteq B_\mathbb{W}(G)$.*

Proof. Suppose first that the hypotheses of Lemma 4.3 hold. We then make the observation that the edges $e \in E^{scc}$ are not used to calculate $B_\mathbb{W}(G)$. Furthermore, every cycle of G is contained in exactly one strongly connected component of this graph. This implies that the Brualdi-type region of the graph is the union of the Brualdi-type regions of its strongly connected components. Therefore, we may, without loss in generality, assume that G consists of a single strongly connected component.

Suppose that both $\gamma = v_1, \ldots, v_m$ and $\delta = v_1, v_m$ are cycles in $C(v_1, G)$ for some $1 < m \leq n$. Note the fact that $\gamma \in C(v_1, G)$ implies that $\gamma' = v_2, \ldots, v_m$ is a cycle in $C(\mathscr{R}_1(G))$, where we let $\mathscr{R}_1(G) = \mathscr{R}_{\bar{v}}(G)$.

From the assumption that v_1 has no adjacent cycles, it follows that $\omega_{mi} = 0$ for $1 < i \leq m$, since otherwise, $\{v_i, v_{i+1}, \ldots, v_m\} \in \mathrm{Adj}(v_1, G)$. Also, since $\gamma \in C(v_1, G) = S(v_1, G)$, we must have $\omega_{i1} = 0$ for $1 < i < m$ as well as $\omega_{mi} = 0$ for $m < i \leq n$, since G is assumed to have one strongly connected component. Therefore,

$$B_\mathbb{W}(G)_\gamma = \{\lambda \in \mathbb{C} : \prod_{i=1}^{m} |\lambda_{ii} L_i| \leq |\omega_{m1} L_m| \prod_{i=1}^{m-1} R_i(G)\}, \tag{4.30}$$

$$B_\mathbb{W}(G)_\delta = \{\lambda \in \mathbb{C} : |\lambda_{11} L_1||\lambda_{mm} L_m| \leq |\omega_{m1} L_m| R_1(G)\}, \tag{4.31}$$

where $R_i(G) = R_i(M(G))$.

Suppose, then, that $\lambda \in B_\mathbb{W}(\mathscr{R}_1(G))_{\gamma'}$. Then

$$|(\lambda_{mm} - \frac{\omega_{m1}\omega_{1m}}{\lambda_{11}}) L_m^1| \prod_{i=2}^{m-1} |\lambda_{ii} L_i| \leq \sum_{i=2}^{m-1} |\frac{\omega_{m1}\omega_{1i}}{\lambda_{11}} L_m^1| \prod_{i=2}^{m-1} R_i(G). \tag{4.32}$$

Here, $L_i^1 = L_i$ for $1 < i < m$, since for each such i, the edge e_{i1} is not in E.

By multiplying both sides of (4.32) by $|q_{11} q_{1m} \lambda_{11}|$, it follows from the triangle inequality that

$$\begin{aligned}
Q_m(G)\Big(|\lambda_{11} L_1 \lambda_{mm} L_m| - |\omega_{1m} L_1 \omega_{m1} L_m|\Big) &\prod_{i=2}^{m-1} |\lambda_{ii} L_i| \\
&\leq Q_m(G)\Big(|\omega_{m1} L_m| R_1(G) - |\omega_{1m} L_1 \omega_{m1} L_m|\Big) \prod_{i=2}^{m-1} R_i(G),
\end{aligned} \tag{4.33}$$

4.3 Brualdi-Type Regions

where $Q_m(G) = Q_m(M(G))$. By use of equation (4.5), we then have

$$B_\mathbb{W}(G)_\delta = \{\lambda \in \mathbb{C} : \prod_{k=1,m} |L_{kk}(q_{kk}\lambda - p_{kk})| \leq \prod_{k=1,m} \left(\sum_{j=1,j\neq k}^{n} |p_{kj}L_{kj}| \right)\}.$$

Hence if $Q_m(G) = 0$, then by calculations analogous to those given in the proof of Theorem 4.3, it follows that $\lambda \in B_\mathbb{W}(G)_\delta$.

Let us now assume that $Q_m(G) \neq 0$. In this case, if $\prod_{i=2}^{m-1} R_i(G) = 0$, it then follows from (4.33) that either $\prod_{i=2}^{m-1} |\lambda_{ii}L_i| = 0$ or $|\lambda_{11}L_1\lambda_{mm}L_m| - |\omega_{m1}L_1\omega_{1m}L_m| = 0$. If the first is the case, then $\lambda \in B_\mathbb{W}(G)_\gamma$. If the latter is the case, then $\lambda \in B_\mathbb{W}(G)_\delta$, since $|\omega_{1m}L_1| \leq R_1(G)$.

If both $\prod_{i=2}^{m-1} R_i(G) \neq 0$ and $|\lambda_{11}L_1\lambda_{mm}L_m| - |\omega_{m1}L_1\omega_{1m}L_m| \neq 0$, then (4.33) implies

$$\frac{\prod_{i=2}^{m-1} |\lambda_{ii}L_i|}{\prod_{i=2}^{m-1} R_i(G)} \leq \frac{|\omega_{m1}L_m|R_1(G) - |\omega_{1m}L_1\omega_{m1}L_m|}{|\lambda_{11}L_1\lambda_{mm}L_m| - |\omega_{m1}L_1\omega_{1m}L_m|}. \tag{4.34}$$

Here, we note that if

$$\frac{|\omega_{m1}L_m|R_1(G) - |\omega_{1m}L_1\omega_{m1}L_m|)}{|\lambda_{11}L_1\lambda_{mm}L_m| - |\omega_{m1}L_1\omega_{1m}L_m|} \leq \frac{|\omega_{m1}L_m|R_1(G)}{|\lambda_{11}L_1\lambda_{mm}L_m|},$$

then it follows from (4.34) together with (4.30) that $\lambda \in B_\mathbb{W}(G)_\gamma$. On the other hand, if this inequality does not hold, then $|\lambda_{11}L_1||\lambda_{mm}L_m| < |\omega_{m1}L_m|R_1(G)$, implying $\lambda \in B_\mathbb{W}(G)_\delta$. Therefore, $B_\mathbb{W}(\mathscr{R}_1(G))_{\gamma'} \subseteq B_\mathbb{W}(G)_\gamma \cup B_\mathbb{W}(G)_\delta \subseteq B_\mathbb{W}(G)$.

Conversely, if $\delta \notin C(G)$, then $\omega_{1m}L_1 = 0$. In this case, equation (4.33) together with (4.30) implies that $B_\mathbb{W}(\mathscr{R}_1(G))_{\gamma'} \subseteq B_\mathbb{W}(G)_\gamma$. Hence, $B_\mathbb{W}(G)_{\gamma'} \subseteq B_\mathbb{W}(G)$, completing the proof. □

We now give a proof of Theorem 4.13.

Proof. First, as in the previous proof, suppose G consists of a single strongly connected component. Moreover, for the vertex $v_1 \in V$, suppose $\text{Adj}(v_1, G) = \emptyset$ and $C(v_1, G) = S(v_1, G)$. Also, let $\gamma' = v_2, \ldots, v_m$ be a cycle in $C(\mathscr{R}_1(G))$ for some $1 < m \leq n$.

Since $\text{Adj}(v_1, G) = \emptyset$, if $\gamma' \in C(G)$, then $M(G, \lambda)_{ij} = M(\mathscr{R}_1(G), \lambda)_{ij}$ for $2 \leq i \leq m$ and $1 \leq j \leq n$, since γ' would otherwise be adjacent to v_1. From this, it follows that $B_\mathbb{W}(\mathscr{R}_1(G))_{\gamma'} = B_\mathbb{W}(G)_{\gamma'} \subseteq B_\mathbb{W}(G)$.

On the other hand, if $\gamma' \notin C(G)$, then at least one edge of the form $e_{i-1,i}$ for $3 \leq i \leq m$ or the edge e_{m2} is not in E. If this is the case, then without loss of generality, assume for notational simplicity that $e_{m2} \notin E$. Furthermore, let

$$\text{Ind}(E) = \{i : e_{i-1,i} \notin E, 3 \leq i \leq m\} \cup \{2\}.$$

We give the set $\text{Ind}(E)$ the ordering $\text{Ind}(E) = \{i_1, \ldots, i_\ell\}$ such that $i_j < i_k$ if and only if $j < k$. Hence, for each $1 \le j \le \ell$, the ordered sets

$$\gamma_j = \{v_1, v_{i_j}, v_{i_j+1}, \ldots, v_{j_\alpha}\} \tag{4.35}$$

are cycles in $C(v_1, G)$, where $j_\alpha = i_{j+1} - 1$ and $\ell_\alpha = m$.

By removing the vertex v_1 from G, it follows from (4.35) that each of the ordered sets

$$\gamma'_j = \{v_{i_j}, v_{i_j+1}, \ldots, v_{j_\alpha}\}$$

is a cycle in $C(\mathcal{R}_1(G))$. Since both $\text{Adj}(v_1, G) = \emptyset$ and $C(v_1, G) = S(v_1, G)$, Lemma 4.3 implies that

$$\bigcup_{j=1}^{\ell} B_W(\mathcal{R}_1(G))_{\gamma'_j} \subseteq B_W(G).$$

The claim, then, is that

$$B_W(\mathcal{R}_1(G))_{\gamma'} \subseteq \bigcup_{j=1}^{\ell} B_W(\mathcal{R}_1(G))_{\gamma'_j}. \tag{4.36}$$

To see this, let $\lambda_{ii}^1 = (\lambda - \omega_{ii} - \dfrac{\omega_{i1}\omega_{1i}}{\lambda_{11}})L_i^1$ and $R_i^1 = \displaystyle\sum_{j=2, j\ne i}^{n} |M(\bar{\mathcal{R}}_1, \lambda)_{ij}|$. Then

$$B_W(\mathcal{R}_1(G))_{\gamma'} = \{\lambda \in \mathbb{C} : \prod_{i=2}^{m} |\lambda_{ii}^1| \le \prod_{i=2}^{m} R_i^1\} \text{ and} \tag{4.37}$$

$$B_W(\mathcal{R}_1(G))_{\gamma'_j} = \{\lambda \in \mathbb{C} : \prod_{i \in \gamma_j} |\lambda_{ii}^1| \le \prod_{i \in \gamma_j} R_i^1\} \text{ for } 1 \le j \le \ell. \tag{4.38}$$

Since the vertex set γ' is the disjoint union of the vertex sets of the cycles γ'_j, the assumption that $\lambda \notin B_W(\mathcal{R}_1(G))_{\gamma'_j}$ for each $1 \le j \le \ell$ implies $\lambda \notin B_W(\mathcal{R}_1(G))_{\gamma'}$ by comparing the product of (4.38) over all $1 \le j \le \ell$ to (4.37). This verifies the claim given in (4.36), which implies that $B_W(\mathcal{R}_1(G))_{\gamma'} \subseteq B_W(G)$.

Since γ' was an arbitrary cycle in $C(\mathcal{R}_1(G))$, it follows that $B_W(\mathcal{R}_1(G)) \subseteq B_W(G)$, completing the proof. □

A proof of Theorem 4.14 is the following.

Proof. If the conditions given in Theorem 4.14 hold for $v = v_1$, both $br_\mathbb{W}(G)$ and $br_\mathbb{W}(\mathscr{R}_1(G))$ exist, since it is assumed that $C_w(G) = \emptyset$ and $C_w(\mathscr{R}_1(G)) = \emptyset$. Moreover, if $S(v_1, G)$ is replaced by $S_{br}(v_1, G)$, and $B_\mathbb{W}(\cdot)$ by $br_\mathbb{W}(\cdot)$, the conclusions of Lemma 4.3 hold by the same argument with the exception that G is not assumed to have a single strongly connected component. Since the same holds for the proof of Theorem 4.13, the result follows. □

4.4 Some Applications

In this section, we discuss some natural applications of isospectral reductions. Our first application deals with estimating the spectra of the Laplacian matrix of a given graph. Following this, we give a method for estimating the spectral radius of a graph using specific types of isospectral reductions. Last, we will use the structure of a given graph to identify particularly types of structural sets that can be used to improve our eigenvalue estimates.

To begin, we note that it is not only possible to reduce a graph G, but it is also possible to reduce both the combinatorial Laplacian matrix and the normalized Laplacian matrix of G. Such matrices are typically defined for undirected graphs without loops or weights, but this definition can be extended to graphs in $\mathbb{G}$ (see Remark 4.2 below). However, here we give the standard definitions of these matrices, since they are of interest in their own right (see [15, 16]).

Let $G = (V, E)$ be an unweighted undirected graph without loops, i.e., a *simple graph*. If G has vertex set $V = \{v_1, \ldots, v_n\}$, and $d(v_i)$ is the degree of vertex v_i, then the *combinatorial Laplacian matrix* $M_L(G)$ of G is given by

$$M_L(G)_{ij} = \begin{cases} d(v_i) & \text{if } i = j, \\ -1 & \text{if } i \neq j \text{ and } v_i \text{ is adjacent to } v_j, \\ 0 & \text{otherwise.} \end{cases}$$

On the other hand, the *normalized Laplacian matrix* $M_\mathscr{L}(G)$ of G is defined as

$$M_\mathscr{L}(G)_{ij} = \begin{cases} 1 & \text{if } i = j \text{ and } d(v_j) \neq 0, \\ \frac{-1}{\sqrt{d(v_i)d(v_j)}} & \text{if } v_i \text{ is adjacent to } v_j, \\ 0 & \text{otherwise.} \end{cases}$$

The interest in the eigenvalues of $M_L(G)$ is that $\sigma(M_L(G))$ gives us structural information regarding the graph G (see, for instance, [15]). Additionally, knowing the eigenvalues $\sigma(M_\mathscr{L}(G))$ is useful in determining the behavior of a number of algorithms on the graph G (see [16]).

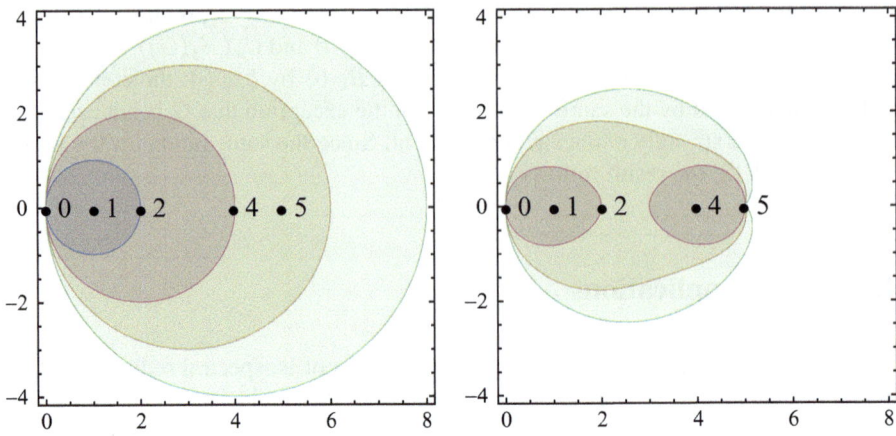

Fig. 4.11 Left: $\Gamma_W(M_L(H))$. Right: $\Gamma_W(\mathscr{R}(M_L(H); S))$, where in each, $\sigma(M_L(H)) = \{0, 1, 2, 4, 5\}$ is indicated

Since $M_L(G)$ and $M_{\mathscr{L}}(G)$ are real matrices, either may be reduced over any subset of their index sets. For example, if $H \in \mathbb{G}_\pi$ is the graph with adjacency matrix

$$M(H) = \begin{bmatrix} 0 & 0 & 0 & 0 & 1 \\ 0 & 0 & 0 & 1 & 1 \\ 0 & 0 & 0 & 1 & 1 \\ 0 & 1 & 1 & 0 & 1 \\ 1 & 1 & 1 & 1 & 0 \end{bmatrix}, \text{ then } M_L(H) = \begin{bmatrix} 1 & 0 & 0 & 0 & -1 \\ 0 & 2 & 0 & -1 & -1 \\ 0 & 0 & 2 & -1 & -1 \\ 0 & -1 & -1 & 3 & -1 \\ -1 & -1 & -1 & -1 & 4 \end{bmatrix}.$$

Reducing $M_L(H)$ over the set $S = \{v_1, v_2, v_3, v_4\}$ yields the matrix

$$\mathscr{R}(M_L(H); S) = \begin{bmatrix} \frac{\lambda-3}{\lambda-4} & \frac{1}{\lambda-4} & \frac{1}{\lambda-4} & \frac{1}{\lambda-4} \\ \frac{1}{\lambda-4} & \frac{2\lambda-7}{\lambda-4} & \frac{1}{\lambda-4} & \frac{-\lambda+5}{\lambda-4} \\ \frac{1}{\lambda-4} & \frac{1}{\lambda-4} & \frac{2\lambda-7}{\lambda-4} & \frac{-\lambda+5}{\lambda-4} \\ \frac{1}{\lambda-4} & \frac{-\lambda+5}{\lambda-4} & \frac{-\lambda+5}{\lambda-4} & \frac{3\lambda-11}{\lambda-4} \end{bmatrix}.$$

Figure 4.11 shows the Gershgorin-type regions of both $M_L(H)$ and $\mathscr{R}(M_L(H); S)$.

Recall that the adjacency matrix of an undirected real graph is symmetric, so its eigenvalues must be real numbers. With this in mind, we note that the Gershgorin-type region associated with an undirected real graph and any of its reductions can be reduced to intervals of the real line.

Remark 4.2. It is possible to generalize the combinatorial Laplacian matrix $M_L(G)$ to any $G \in \mathbb{G}^n$ if G has no loops by setting $M_L(G)_{ij} = -M(G)_{ij}$ for $i \neq j$ and

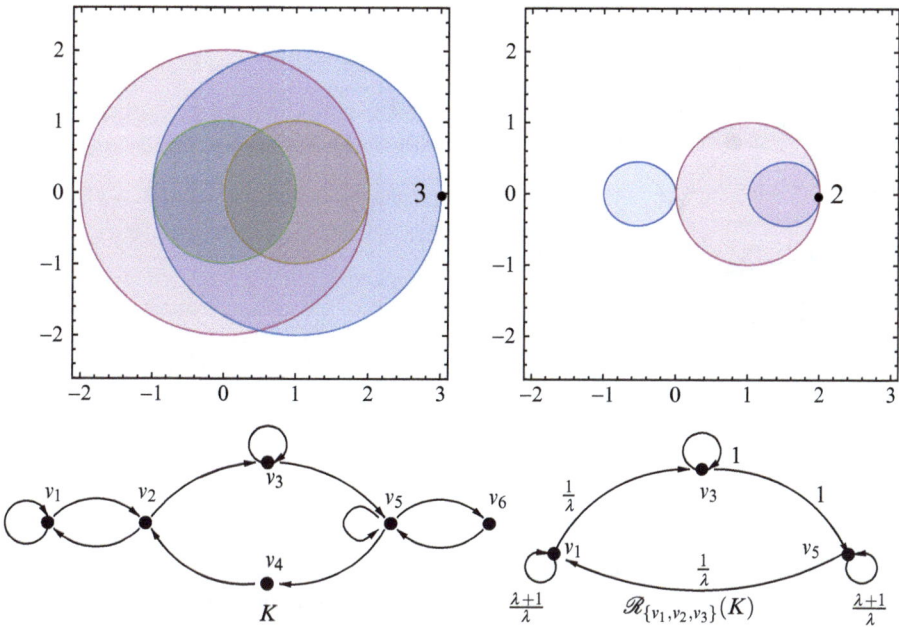

Fig. 4.12 *Top Left*: The region $\Gamma_{\mathbb{W}}(K)$ from which $\rho(K) \leq 3$. *Top Right*: The region $\Gamma_{\mathbb{W}}(\mathscr{R}_{\{v_1,v_2,v_3\}}(K))$ from which $\rho(K) \leq 2$

$M_L(G)_{ii} = \sum_{j=1, j\neq i}^{n} M(G)_{ij}$. This generalization is consistent, for example, with what is done for weighted digraphs in [31].

We now turn our attention to estimating the spectral radius of a graph, or equivalently, a matrix. Recall that for a graph $G \in \mathbb{G}_\pi$, the spectral radius of G, denoted by $\rho(G)$, is the maximum among the absolute values of the elements in $\sigma(G)$, i.e.,

$$\rho(G) = \max_{\ell \in \sigma(G)} |\ell|.$$

If a graph $G \in \mathbb{G}_\pi$ has a complete structural set S, then by Corollary 2.5, the spectral radius $\rho(G)$ is equal to $\rho(\mathscr{R}_S(G))$. For example, in the graph K shown in Fig. 4.12, the vertices v_2, v_4, v_6 are the vertices of K without loops. Since $\{v_1, v_3, v_5\} \in st(K)$, these vertices form a complete structural set of K, from which it follows that $\rho(K) = \rho(\mathscr{R}_{\{v_1,v_3,v_5\}}(K))$.

By calculating the region $\Gamma_{\mathbb{W}}(K)$, which is shown in Fig. 4.12 (top left), we can then estimate that $\rho(K) \leq 3$. However, using the Gershgorin-type region of the reduced graph $\Gamma_{\mathbb{W}}(\mathscr{R}_{\{v_1,v_3,v_5\}}(K))$, shown in Fig. 4.12 (top right), our estimate of the graphs spectral radius improves to $\rho(K) \leq 2$.

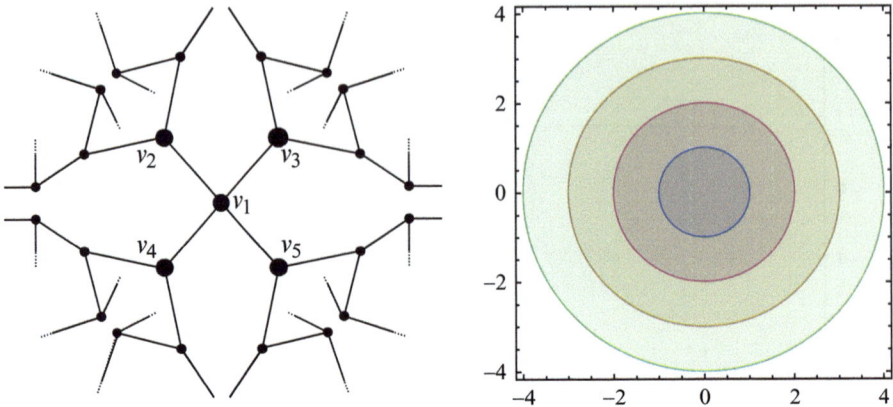

Fig. 4.13 The graph G (*left*) and its associated Gershgorin-type region $\Gamma_{\mathbb{W}}(G)$ (*right*) considered in Example 4.11

It should be noted that a graph $G \in \mathbb{G}_\pi$ may have more than one complete structural set. In this case, there may be many ways to reduce G while maintaining its spectral radius.

As a final application of the theory developed in this chapter, we consider reducing graphs over specific structural sets. Our goal is to improve our eigenvalue estimates when some structural feature of the graph is known.

Suppose $G = (V, E, \omega)$ is a graph in $\mathbb{G}_\pi^n$. If the sets $\Gamma_{\mathbb{W}}(G)_i$ for $1 \leq i \leq n$ are known or can be estimated by some structural knowledge of G, then it is possible to decide over which structural sets to reduce. That is, it may be possible to identify structural sets $U \subset V$ such that $v_i \notin U$ and

$$\partial \Gamma_{\mathbb{W}}(G)_i \nsubseteq \bigcup_{j \neq i} \Gamma_{\mathbb{W}}(G)_j.$$

If this can be done, Theorem 4.4 implies that a strictly better estimate of $\sigma(G)$ can be achieved by reducing over U.

Example 4.11. Consider the graph $G = (V, E, \omega)$ in Fig. 4.13 (left), where $G \in \mathbb{G}^n$ for some $n \geq 5$. Suppose it is known that G is an unweighted undirected graph, in which $d(v_1) = 4$, $d(v_2) = d(v_3) = d(v_4) = d(v_5) = 3$, and $d(v_i) \in \{0, 1, 2, 3\}$ for all $6 \leq i \leq n$. Then the sets $\Gamma_{\mathbb{W}}(G)_i$ are each disks centered at the origin of radius $r \in \{0, 1, 2, 3, 4\}$, shown in Fig. 4.13 (right).

Since $\partial \Gamma_{\mathbb{W}}(M(G))_1 = \{\lambda \in \mathbb{C} : |\lambda| = 4\}$ is an infinite set of points and

$$\partial \Gamma_{\mathbb{W}}(M(G))_1 \nsubseteq \bigcup_{i=2}^{n} \Gamma_{\mathbb{W}}(M(G))_i,$$

4.4 Some Applications

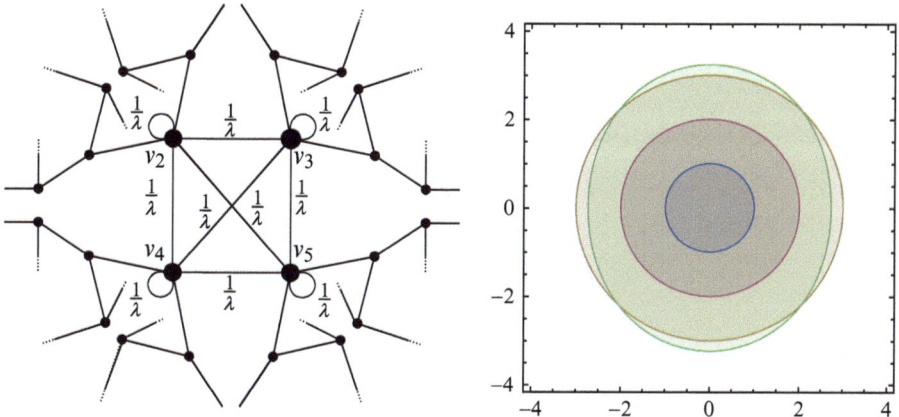

Fig. 4.14 The reduced graph $\mathscr{R}_S(G)$ (*left*) and its associated Gershgorin-type region $\Gamma_W(\mathscr{R}_S(G))$ (*right*) considered in Example 4.11

then Theorem 4.4 implies the following. If $S = V - \{v_1\}$, the graph $\mathscr{R}_S(G)$ has a strictly smaller Gershgorin-type region than G, as can be seen in Fig. 4.14.

Considering the fact that n may be quite large, this example is intended to illustrate that eigenvalue estimates can be improved with a minimal amount of effort if some simple structural feature(s) of the graph are known.

Observe that as a matrix $M \in \mathbb{C}^{n \times n}$ is reduced, its entries may contain increasingly larger powers of λ. Therefore, the more M is reduced, the more complicated it can become to compute its Gershgorin-, Brauer-, or Brualdi-type region.

Fortunately, there is a fairly simple bound for how large these powers of λ can become. If $S \subset N$, let $\mathscr{R}(M; S)_{ij} = p_{ij}/q_{ij}$, where $p_{ij}, q_{ij} \in \mathbb{C}[\lambda]$. Then Lemma 1.1 can be used to show that

$$deg(p_{ij}) \leq deg(q_{ij}) \leq |\bar{S}_1| < n.$$

For instance, in Example 4.4, the matrix $M_0 \in \mathbb{C}^{5 \times 5}$ is reduced over $S_1 = \{1, 2, 3\}$. The result is the matrix

$$M_1 = \begin{bmatrix} \frac{\lambda+1}{\lambda^2} & \frac{1}{\lambda} & \frac{\lambda+1}{\lambda} \\ \frac{2\lambda+1}{\lambda^2} & \frac{1}{\lambda} & \frac{1}{\lambda} \\ 0 & 1 & 0 \end{bmatrix}. \quad (4.39)$$

As can be seen, the largest power of λ in any entry of M_1 is $2 = |\bar{S}|$.

Therefore, the eigenvalue regions associated with a matrix may become harder to compute as a matrix is reduced, but only marginally so. For both Gershgorin- and Brauer-type regions, this is offset by the fact that there are fewer individual regions to compute once the matrix has been reduced.

Chapter 5
Pseudospectra and Inverse Pseudospectra

The pseudospectrum of a complex-valued matrix $A \in \mathbb{C}^{n \times n}$ is the collection of scalars that behave, to within a given tolerance, like an eigenvalue of A. In this chapter, we extend the definition of pseudospectrum to matrices with entries that are rational functions and introduce the notion of a matrix's inverse pseudospectrum.

One of the main results we prove is that for a given tolerance, the pseudospectrum of a reduced matrix is always contained in the pseudospectrum of the original matrix. As a consequence, the eigenvalues of a reduced matrix are less susceptible to perturbations than those that correspond to the matrix itself. This is important, for instance, in systems that correspond to isospectrally reduced matrices. Our prime example of such systems is that of linear mass–spring networks in which there is limited access.

The reason we study mass–spring networks is that such systems allow us to give a physical interpretation of the concepts of pseudospectra and inverse pseudospectra introduced in this chapter. The major result, that the pseudospectrum of a network shrinks under reduction, also has a physical meaning in this context.

5.1 Pseudospectra

We begin by giving the definition of the pseudospectra of a matrix $A \in \mathbb{C}^{n \times n}$. In fact, we give three equivalent definitions. To do so, we first define the notion of a compatible matrix and vector norm.

A matrix norm $\|\cdot\|$ on $\mathbb{C}^{n \times n}$ is compatible with a vector norm $\|\cdot\|'$ on $\mathbb{C}^n$ if

$$\|A\mathbf{v}\|' \leq \|A\| \|\mathbf{v}\|' \text{ for all } A \in \mathbb{C}^{n \times n} \text{ and } \mathbf{v} \in \mathbb{C}^n.$$

We now define the pseudospectra of a matrix $A \in \mathbb{C}^{n \times n}$.

Definition 5.1. Let $\epsilon > 0$. The ϵ-*pseudospectrum* of $A \in \mathbb{C}^{n \times n}$ is defined equivalently by the following:

(a) *Eigenvalue perturbation*:

$$\sigma_\epsilon(A) = \{\lambda \in \mathbb{C} : ||(A - \lambda I)\mathbf{v}|| < \epsilon \text{ for some } \mathbf{v} \in \mathbb{C}^n \text{ with } ||\mathbf{v}|| = 1\};$$

(b) *The resolvent*:

$$\sigma_\epsilon(A) = \{\lambda \in \mathbb{C} : ||(A - \lambda I)^{-1}|| > \epsilon^{-1}\} \cup \sigma(A); \text{ and}$$

(c) *Perturbation of the matrix*:

$$\sigma_\epsilon(A) = \{\lambda \in \mathbb{C} : \lambda \in \sigma(A + E) \text{ for some } E \in \mathbb{C}^{n \times n} \text{ with } ||E|| < \epsilon\}.$$

The matrix and vector norms in (a)–(c) are assumed to be compatible.

A proof that (a)–(c) of Definition 5.1 define the same region can be found in [26].

We note that the assumption $||\mathbf{v}|| = 1$ in part (a) of Definition 5.1 is necessary, since any given $\lambda \in \mathbb{C}$ could be in $\sigma_\epsilon(A)$ by choosing $||\mathbf{v}||$ small enough. Also, having $\sigma(A)$ on the right-hand side of part (*b*) is necessary, since the pseudospectrum $\sigma_\epsilon(A)$ in parts (a) and (c) contain $\sigma(A)$, but the matrix $A - \lambda I$ in part (b) is noninvertible for every $\lambda \in \sigma(A)$.

Example 5.1. Consider the matrix $A \in \mathbb{C}^{n \times n}$ with (0, 1)-entries given by

$$A = \begin{bmatrix} 0 & 0 & 1 & 1 & 0 & 0 \\ 0 & 1 & 0 & 0 & 1 & 1 \\ 1 & 0 & 1 & 0 & 0 & 0 \\ 0 & 1 & 0 & 1 & 0 & 0 \\ 1 & 0 & 0 & 0 & 0 & 0 \\ 0 & 1 & 0 & 0 & 0 & 0 \end{bmatrix},$$

considered in Example 1.3. The pseudospectra $\sigma_\epsilon(A)$, for $\epsilon = 1, 1/2$, and $1/4$, are shown in Fig. 5.1 (left). The eigenvalues $\sigma(M) = \{2, -1, 1, 1, 0, 0\}$ are also indicated.

For $A \in \mathbb{C}^{n \times n}$, suppose that for a given *tolerance* $\epsilon > 0$, there exist a scalar $\lambda \in \mathbb{C}$ and a unit vector $\mathbf{v} \in \mathbb{C}^n$ for which $||(A - \lambda I)\mathbf{v}|| < \epsilon$. If this is the case, then the vector $\mathbf{v}$ is said to be an ϵ-*pseudoeigenvector* of the matrix A corresponding to the ϵ-*pseudoeigenvalue* λ. Definition 5.1 (a) states that the ϵ-pseudospectrum of A is defined as the set of all such λ that have an associated ϵ-pseudoeigenvector.

To extend the notion of an ϵ-pseudospectrum to a matrix $M(\lambda) \in \mathbb{W}_\pi^{n \times n}$, we first need to have an idea of what it means to be an eigenvector of $M(\lambda)$. If $\lambda \in \sigma(A)$, where $A \in \mathbb{C}^{n \times n}$, then there is always at least one eigenvector $\mathbf{v} \in \mathbb{C}^n$ of A associated with λ. However, recall from Chap. 1 that a matrix $M(\lambda) \in \mathbb{W}_\pi^{n \times n}$

5.1 Pseudospectra

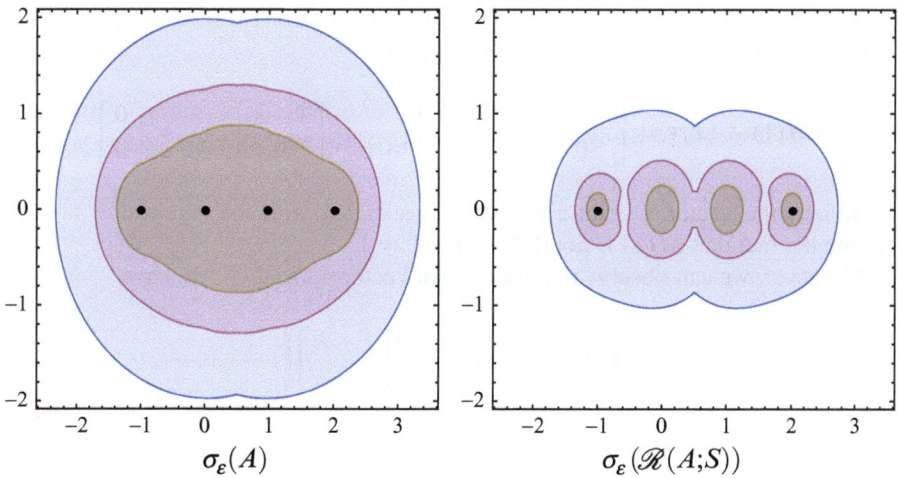

Fig. 5.1 Pseudospectra of the matrices given in Example 5.3 for $\epsilon = 1$ (*blue*), $\epsilon = 1/2$ (*red*), and $\epsilon = 1/4$ (*tan*), obtained using the matrix 1-norm. The spectra $\sigma(A) = \{2, -1, 1, 1, 0, 0\}$ and $\sigma(\mathscr{R}(A; S)) = \{2, -1\}$ are shown as black dots.

may have an eigenvalue λ_0 at which the matrix $M(\lambda_0)$ is undefined. This may seem problematic, especially if we wish to associate an eigenvector with $\lambda_0 \in \sigma(M)$. As it turns out, we can always associate an eigenvector $\mathbf{v} \in \mathbb{C}^n$ with each eigenvalue of $M(\lambda)$.

Suppose λ_0 is a solution of the equation $\det(M(\lambda) - \lambda I) = 0$. Then the standard theory of linear algebra implies that there is a vector $\mathbf{v}$ such that $(M(\lambda) - \lambda I)\mathbf{v} = 0$ when this product is evaluated at $\lambda = \lambda_0$. Keeping this sequence in mind of first multiplying then evaluating, we define the product of a matrix and vector as follows. For every $M(\lambda) \in \mathbb{W}_\pi^{n \times n}$, $\mathbf{v} \in \mathbb{C}^n$, and $\lambda \in \mathbb{C}$, we let the matrix/vector product be given by

$$(M(\lambda) - \lambda I)\mathbf{v} \equiv [(M(s) - sI)\mathbf{v}]|_{s=\lambda}. \tag{5.1}$$

This definition allows us to associate an eigenvector to each eigenvalue of a matrix $M(\lambda) \in \mathbb{W}_\pi^{n \times n}$. To give an explicit demonstration of this procedure, we consider the following example.

Example 5.2. Consider the matrix $M(\lambda) \in \mathbb{W}^{2 \times 2}$ given by

$$M(\lambda) = \begin{bmatrix} 1 & \frac{1}{\lambda - 1} \\ 0 & 1 \end{bmatrix}.$$

Here, one can readily see that $\sigma(M) = \{1, 1\}$ but $1 \notin dom(M)$. Although $M(1)$ is undefined, the vector $\mathbf{v} = [1\ 0]^T$ has the property

$$(M(1) - 1I)\mathbf{v} = \begin{bmatrix} 1-s & \frac{1}{s-1} \\ 0 & 1-s \end{bmatrix} \begin{bmatrix} 1 \\ 0 \end{bmatrix} \bigg|_{s=1} = \begin{bmatrix} 1-s \\ 0 \end{bmatrix} \bigg|_{s=1} = \begin{bmatrix} 0 \\ 0 \end{bmatrix}.$$

Therefore, the vector $\mathbf{v}$ is an eigenvector associated with the eigenvalue $1 \in \mathbb{C}$ despite the fact that $M(\lambda)$ is not defined for $\lambda = 1$.

Moreover, we can observe that for a given vector norm $||\cdot||$, we have

$$||(M(\lambda) - \lambda I)\mathbf{v}|| = \left\|\begin{bmatrix} 1-\lambda \\ 0 \end{bmatrix}\right\|.$$

Hence, the size of $(M(\lambda) - \lambda I)\mathbf{v}$ varies continuously with respect to λ even where $M(\lambda)$ is undefined. This type of continuity can be shown to hold in general for each matrix $M(\lambda) \in \mathbb{W}_\pi^{n \times n}$ near $\sigma(M)$, and it is essential in describing the pseudospectra of this class of matrices.

We are now in a position to define the ϵ-pseudospectrum of a matrix $M(\lambda) \in \mathbb{W}_\pi^{n \times n}$. To do this, we let $cl(\Omega)$ denote the closure of Ω in $\mathbb{C}$ for the set $\Omega \subset \mathbb{C}$.

Definition 5.2. Let $\epsilon > 0$. The ϵ-*pseudospectrum* of $M(\lambda) \in \mathbb{W}_\pi^{n \times n}$ is defined equivalently by the following conditions:

(a) *Eigenvalue perturbation*:

$$\sigma_\epsilon(M) = cl\big(\{\lambda \in \mathbb{C} : ||(M(\lambda) - \lambda I)\mathbf{v}|| < \epsilon \text{ for some } \mathbf{v} \in \mathbb{C}^n \text{ with } ||\mathbf{v}|| = 1\}\big).$$

(b) *The resolvent*:

$$\sigma_\epsilon(M) = cl\big(\{\lambda \in \mathbb{C} : ||(M(\lambda) - \lambda I)^{-1}|| > \epsilon^{-1}\}\big).$$

(c) *Perturbation of the matrix*:

$$\sigma_\epsilon(M) = cl\big(\{\lambda \in \mathbb{C} : \lambda \in \sigma(M(\lambda) + E) \text{ for some } E \in \mathbb{C}^{n \times n} \text{ with } ||E|| < \epsilon\}\big).$$

We assume that the matrix and vector norms in (a)–(c) are compatible.

An immediate consequence of Definition 5.2 is that the eigenvalues of a matrix $M \in \mathbb{W}_\pi^{n \times n}$ belong to its pseudospectra:

$$\sigma(M) \subset \sigma_\epsilon(M) \text{ for each } \epsilon > 0.$$

The proof that Definitions 5.2(a)–(c) are equivalent relies on the proof that Definitions 5.2(a)–(c) are equivalent for scalar-valued matrices. For completeness,

5.1 Pseudospectra

Fig. 5.2 The mass–spring network of Example 5.4 with boundary nodes $S = \{1, 4\}$ and interior nodes $\bar{S} = \{2, 3\}$

the proof that Definitions 5.2(a)–(c) are equivalent is included at the end of the chapter, in Sect. 5.4.

Remark 5.1. In Definition 5.2(c), we could have defined 511.5

$$\sigma_\epsilon(M) = cl\big(\{\lambda \in \mathbb{C} : \lambda \in \sigma(M(\lambda) + E(\lambda)) \text{ for some } E \in \mathbb{W}^{n \times n} \text{ with } \|E(\lambda)\| < \epsilon\}\big),$$

so that $M(\lambda)$ is perturbed by the matrix $E(\lambda)$ of rational functions. However, for a fixed $\lambda \in dom(E)$, the matrix $E(\lambda)$ is in $\mathbb{C}^{n \times n}$, so that this alternative definition is equivalent to the one given above, which is simpler.

To give an example of the ϵ-pseudospectrum of a matrix of rational functions, we consider the following.

Example 5.3. Let A be the matrix given in Example 5.1 and $S = \{1, 2\}$. Then

$$\mathscr{R}(A; S) = \begin{bmatrix} \frac{1}{\lambda-1} & \frac{1}{\lambda-1} \\ \frac{1}{\lambda} & \frac{\lambda+1}{\lambda} \end{bmatrix} \in \mathbb{W}_\pi^{2 \times n}.$$

The pseudospectra of both A and $\mathscr{R}(A; S)$ are displayed in Fig. 5.1 for $\epsilon = 1, 1/2, 1/4$ using the matrix 1-norm. Notice that although $0, 1 \in \sigma(A)$, these values do not belong to $\sigma(\mathscr{R}(A; S))$, because of cancellations resulting from the matrix reduction, i.e., $A_{\bar{S}\bar{S}} = \{0, 0, 1, 1\}$. However, for the ϵ we consider, $0, 1 \in \sigma_\epsilon(\mathscr{R}(A; S))$, meaning that these eigenvalues remain as pseudoeigenvalues of the reduced matrix.

In the previous chapters we described a number of ways of using isospectral matrix and graph reductions. In Chap. 2, graph reductions were used to simplify the structure of a network while maintaining its spectral properties. In Chap. 4, matrix reductions were used to gain improved eigenvalue estimates. Here, we use isospectral reductions to study the dynamics of mass–spring networks in which there is limited access.

Example 5.4. Consider the mass–spring network illustrated in Fig. 5.2, with nodes at locations x_i, $i = 1, 2, 3, 4$, lying on a line and springs linking nearest neighbors. For simplicity, we assume that all the springs have the same spring constant $k = 1$ and that all the nodes have unit mass. (The precise position of the nodes on the line does not matter for this discussion.)

Suppose each node x_i is subject to a time-harmonic displacement $u_i(\omega)e^{j\omega t}$ with frequency ω in the direction of the line and $j = \sqrt{-1}$. Then the resulting force at node x_i is also time-harmonic in the direction of the line and is of the form $f_i(\omega)e^{j\omega t}$. Writing the balance of forces acting on each node with the laws of motion, one can show that the vector of forces $\mathbf{f}(\omega) = [f_1(\omega), \ldots, f_4(\omega)]^T$ is linearly related to the vector of displacements $\mathbf{u}(\omega) = [u_1(\omega), \ldots, u_4(\omega)]^T$ by the equation

$$\mathbf{f}(\omega) = (K - \omega^2 I)\mathbf{u}(\omega). \tag{5.2}$$

Here the matrix K is the *stiffness* matrix

$$K = \begin{bmatrix} 1 & -1 & 0 & 0 \\ -1 & 2 & -1 & 0 \\ 0 & -1 & 2 & -1 \\ 0 & 0 & -1 & 1 \end{bmatrix}.$$

If we let $\lambda \equiv \omega^2$, we see that the eigenmodes $\sigma(K) = \{2 \pm \sqrt{2}, 2, 0\}$ of the stiffness matrix K correspond to nonzero displacements that do not generate forces on the network nodes. For instance, the eigenmode corresponding to the zero frequency is $\mathbf{u} = [1, 1, 1, 1]^T$; i.e., by displacing all nodes by the same amount, there are no net forces at the nodes.

Since the eigenvalues of K correspond to frequencies for which there exists a nonzero displacement that generates no forces on these nodes, the pseudoeigenvalues of this system have a similar physical interpretation. Namely, the pseudospectra indicate the frequencies for which there is a displacement that generates "small" forces relative to the (norm of the) displacement.

For example, the frequency $\omega^2 = 2.1$ in Fig. 5.3 (left) is within the tan tolerance region for $\epsilon = 1/4$. Hence, there is a nonzero vector of displacements such that the forces generated from this displacement have norm equal to $\epsilon = 1/4$ times the norm of this displacement vector.

Now suppose that we have access only to certain boundary nodes of this network, say $S = \{1, 4\}$. Then we can write the equilibrium of forces at the interior nodes $\bar{S} = \{2, 3\}$ and conclude that the net forces $\mathbf{f}_S$ at the boundary nodes S depend linearly on the displacements $\mathbf{u}_S$ at the terminal nodes according to the equation

$$\mathbf{f}_S(\omega) = (\mathscr{R}_{\omega^2}(K; S) - \omega^2 I)\mathbf{u}_S(\omega). \tag{5.3}$$

The spectrum and inverse spectrum of the network response are then

$$\sigma(\mathscr{R}_{\omega^2}(K; S)) = \{2 \pm \sqrt{2}, 2, 0\} \text{ and } \sigma^{-1}(\mathscr{R}_{\omega^2}(K; S)) = \{3, 1\}.$$

The eigenvalues of $\mathscr{R}_{\omega^2}(K; S)$ correspond to frequencies for which there is a displacement of the boundary nodes S that generates no forces at these nodes. Conversely, the inverse eigenvalues of $\mathscr{R}_{\omega^2}(K; S)$ correspond to frequencies at which there is a displacement of the boundary nodes for which the resulting forces are infinitely large.

5.1 Pseudospectra

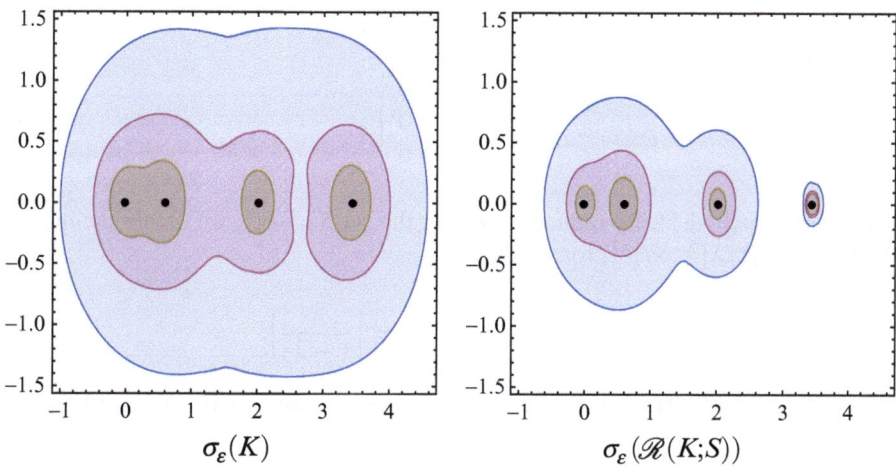

Fig. 5.3 Pseudospectra of the stiffness matrix K for the mass–spring system and its reduction $\mathscr{R}_\lambda(K, S)$ from Example 5.4. The latter corresponds to the effective stiffness of the mass–spring system when we have access only to nodes $S = \{1, 4\}$. The tolerances shown are $\epsilon = 1$ (*blue*), $\epsilon = 1/2$ (*red*), and $\epsilon = 1/4$ (*tan*), using the matrix 1-norm. The spectra of the respective matrices are indicated

That is, if we have access only to the boundary nodes $S = \{1, 4\}$, then the pseudoeigenvalues of $\mathscr{R}_{\omega^2}(K; S)$ correspond to frequencies for which there is a displacement at the boundary nodes S that generates very small forces on these nodes. The pseudospectra regions of $\mathscr{R}_\lambda(K; S)$ are shown in Fig. 5.3 (right) for $\epsilon = 1, 1/2, 1/4$.

Observe that the pseudospectra of $\mathscr{R}_\lambda(K; S)$ are included in the pseudospectra of K for a given tolerance ϵ. This implies that less access to network nodes leads to fewer frequencies for which displacements generate relatively small forces. Phrased less formally, the more a network is reduced, the less susceptible to perturbations its eigenvalues are.

In the standard theory of pseudospectra, if $A \in \mathbb{C}^{n \times n}$ has complex entries, then its pseudospectrum $\sigma_\epsilon(A)$ can have at most n connected components, each corresponding to at least one eigenvalue of A. In contrast, a matrix $M(\lambda) \in \mathbb{W}_\pi^{n \times n}$ can have more than n eigenvalues. Hence, $\sigma_\epsilon(M)$ can have more than n connected components. This is why the region $\sigma_{1/4}(\mathscr{R}(K; \{1, 4\}))$, shown in Fig. 5.3 (right), can have four connected components, although $\mathscr{R}(K; \{1, 4\})$ is a 2×2 matrix.

Before ending this section, we note that in Examples 5.3 and 5.4, we have both $\sigma(A) \subset \sigma_\epsilon(\mathscr{R}(A; S))$ and $\sigma(K) \subset \sigma_\epsilon(\mathscr{R}(K; S))$ for the ϵ we consider. Based on these examples, it seems that even under reduction, the ϵ-pseudospectrum of the reduced matrix "remembers" where the eigenvalues of the original matrix are. However, this is not always the case, as the following example shows.

Example 5.5. Consider the matrix $M \in \mathbb{C}^{3\times 3}$ given by

$$M = \begin{bmatrix} 0 & 1 & 0 \\ 1 & 0 & 0 \\ 0 & 1 & 0 \end{bmatrix},$$

with $\sigma(M) = \{0, \pm 1\}$. By reducing M over the set $S = \{1\}$, we obtain the matrix $\mathscr{R}(M; S) = [1/\lambda] \in \mathbb{W}_\pi^{1\times 1}$, for which

$$\|(\mathscr{R}(M; S) - \lambda I)^{-1}\| = \left|\frac{\lambda}{1-\lambda^2}\right|.$$

Hence, $0 \notin \sigma_\epsilon(\mathscr{R}(M; S))$ for every ϵ. Moreover, since $\sigma(M_{\bar{S}\bar{S}}) = \{0, 0\}$ for $\bar{S} = \{2, 3\}$, it is not always the case that either $\sigma(M)$ or $\sigma(M_{\bar{S}\bar{S}})$ is contained in $\sigma_\epsilon(\mathscr{R}(M; S))$. Therefore, a matrix does not necessarily even approximately remember its eigenvalues under reduction.

5.2 Pseudospectra Under Isospectral Reduction

In this section, we investigate how the pseudospectra of a matrix $M \in \mathbb{W}_\pi^{n\times n}$ is affected by an isospectral reduction. In order to study this change in pseudospectra, we need to consider two vector norms. Specifically, we need one norm $\|\cdot\|$ defined on $\mathbb{C}^n$ for the pseudospectrum of M and another norm $\|\cdot\|'$ defined on $\mathbb{C}^{|S|}$ ($0 < |S| < n$) for the pseudospectrum of $\mathscr{R}(M; S)$. Our comparison of the pseudospectra of the original and reduced matrices assumes that for each vector $\mathbf{v} = (\mathbf{v}_S^T, \mathbf{v}_{\bar{S}}^T)^T \in \mathbb{C}^n$, these two norms are related by

$$\|\mathbf{v}\| = \left\|\begin{bmatrix} \mathbf{v}_S \\ \mathbf{v}_{\bar{S}} \end{bmatrix}\right\| \geq \left\|\begin{bmatrix} \mathbf{v}_S \\ 0 \end{bmatrix}\right\| = \|\mathbf{v}_S\|'. \tag{5.4}$$

Examples of norms satisfying property (5.4) are the p−norms for $1 \leq p \leq \infty$. For the sake of simplicity, we use the same notation for both of these $\mathbb{C}^n$ and $\mathbb{C}^{|S|}$ norms.

The following theorem describes how the ϵ-pseudospectrum of a matrix $M(\lambda)$ is related to the ϵ-pseudospectrum of the isospectral reduction $\mathscr{R}_\lambda(M; S)$. It says that the ϵ-pseudospectra of the reduced matrix is contained in the ϵ-pseudospectra of the original matrix, for each $\epsilon > 0$.

Theorem 5.1. *For $M(\lambda) \in \mathbb{W}_\pi^{n\times n}$, let $S \subset N$. Then $\sigma_\epsilon(\mathscr{R}(M; S)) \subseteq \sigma_\epsilon(M)$ for every $\epsilon > 0$, provided the $\mathbb{C}^n$ and $\mathbb{C}^{|S|}$ norms in the pseudospectra definitions satisfy (5.4).*

Proof. For $M(\lambda) \in \mathbb{W}_\pi^{n\times n}$, let S and $\bar{S}$ form a nonempty partition of N. We assume, without loss of generality, that for the vector $\mathbf{v} \in \mathbb{C}^n$, we have $\mathbf{v} = (\mathbf{v}_S^T, \mathbf{v}_{\bar{S}}^T)^T$.

5.2 Pseudospectra Under Isospectral Reduction

Moreover, for $\tilde{\lambda}_0 \in \mathbb{C}$ and $\epsilon > 0$, suppose $\mathbf{v}_S \in \mathbb{C}^{|S|}$ is a unit vector such that

$$\|(\mathscr{R}(M; S) - \tilde{\lambda}_0 I)\mathbf{v}_S\| < \epsilon. \tag{5.5}$$

Since $\sigma(M_{\bar{S}\bar{S}})$ and $\overline{\text{dom}(M)}$ are finite sets, then by continuity, there is a neighborhood U of $\tilde{\lambda}_0$ such that

(i) $M(\lambda) \in \mathbb{C}^{n \times n}$ for $\lambda \in U - \{\tilde{\lambda}_0\}$;
(ii) $\sigma(M_{\bar{S}\bar{S}}) \cap (U - \{\tilde{\lambda}_0\}) = \emptyset$; and
(iii) $\|(\mathscr{R}(M; S) - \lambda I)\mathbf{v}_S\| < \epsilon$ for $\lambda \in U - \{\tilde{\lambda}_0\}$.

Observe that for each $\lambda_0 \in U - \{\tilde{\lambda}_0\}$, it follows that the vector

$$\mathbf{v}_{\bar{S}} = -(M(\lambda_0)_{\bar{S}\bar{S}} - \lambda_0 I)^{-1} M(\lambda_0)_{\bar{S}S} \mathbf{v}_S$$

is defined. Since $\mathbf{v} = (\mathbf{v}_S^T, \mathbf{v}_{\bar{S}}^T)^T$, we have

$$(M(\lambda_0) - \lambda_0 I)\mathbf{v} = \begin{bmatrix} (M - \lambda I)_{SS}\mathbf{v}_S + (M - \lambda I)_{S\bar{S}}\mathbf{v}_{\bar{S}} \\ (M - \lambda I)_{\bar{S}S}\mathbf{v}_S + (M - \lambda I)_{\bar{S}\bar{S}}\mathbf{v}_{\bar{S}} \end{bmatrix}\bigg|_{\lambda=\lambda_0}$$

$$= \begin{bmatrix} M_{SS}\mathbf{v}_S - M_{S\bar{S}}(M_{\bar{S}\bar{S}} - \lambda I)^{-1} M_{\bar{S}S}\mathbf{v}_S \\ M_{\bar{S}S}\mathbf{v}_S - (M_{\bar{S}\bar{S}} - \lambda I)(M_{\bar{S}\bar{S}} - \lambda I)^{-1} M_{\bar{S}S}\mathbf{v}_S \end{bmatrix}\bigg|_{\lambda=\lambda_0}$$

$$= \begin{bmatrix} (\mathscr{R}(M; S) - \lambda I)\mathbf{v}_S \\ 0 \end{bmatrix}\bigg|_{\lambda=\lambda_0}.$$

By the property (5.4), regarding the norms in $\mathbb{C}^n$ and $\mathbb{C}^{|S|}$, we have

$$\|(M(\lambda_0) - \lambda_0 I)\mathbf{v}\| = \|(\mathscr{R}(M(\lambda_0); S) - \lambda_0 I)\mathbf{v}_S\| < \epsilon. \tag{5.6}$$

Since $\mathbf{v}_S \neq \mathbf{0}$, consider the unit vector $\mathbf{u} = \mathbf{v}/\|\mathbf{v}\| \in \mathbb{C}^n$. Again by (5.4), we have $\|\mathbf{v}\| \geq \|\mathbf{v}_S\| = 1$. Therefore, we obtain the bound

$$\|(M(\lambda_0) - \lambda_0 I)\mathbf{u}\| = \frac{\|(M(\lambda_0) - \lambda_0 I)\mathbf{v}\|}{\|\mathbf{v}\|} \leq \|(M(\lambda_0) - \lambda_0 I)\mathbf{v}\| < \epsilon,$$

where the last inequality comes from (5.6). This implies $\lambda_0 \in \sigma_\epsilon(M)$.

Since this holds for every $\lambda_0 \in U - \{\tilde{\lambda}_0\}$, we must have $\tilde{\lambda}_0 \in cl(\sigma_\epsilon(M))$. Since $\sigma_\epsilon(M)$ is a closed set, it follows that $\tilde{\lambda}_0 \in \sigma_\epsilon(M)$. Since $\tilde{\lambda}_0$ is an arbitrary point in $\sigma_\epsilon(\mathscr{R}(M; S))$, the result follows by inequality (5.5). □

Example 5.6. For the mass–spring network introduced in Example 5.4, we consider four different sets of boundary nodes $\{1, 2, 3, 4\} \supset \{1, 2, 4\} \supset \{1, 4\} \supset \{1\}$. Note that Theorem 5.1 implies that the corresponding pseudospectra for a given ϵ obey the same inclusions. This is shown in Fig. 5.4 for $\epsilon = 1, 1/2, 1/4$.

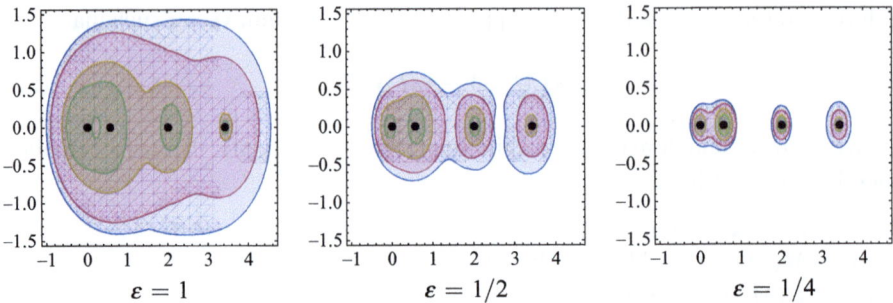

Fig. 5.4 For $\epsilon = 1, 1/2, 1/4$, the pseudospectra of the matrix K from the mass–spring system in Example 5.4 is shown (*blue*) together with the pseudospectra for the reduced matrices with terminal nodes $S = \{1, 2, 4\}$ (*red*), $S = \{1, 4\}$ (*tan*), and $S = \{1\}$ (*green*). Again the matrix 1-norm is used, and the eigenvalues $\sigma(K) = \{0, 2, 2 \pm \sqrt{2}\}$ are indicated. Note how the pseudospectra shrink as the number of boundary nodes decreases

In terms of the mass–spring network, this means that as we increase the number of internal degrees of freedom (or decrease the number of boundary nodes), it becomes harder to find frequencies for which there is a displacement that generates forces of magnitude below a certain fixed level. Hence the fewer boundary nodes we have, the more robust the frequencies are that generate small forces.

Notice that the inclusion given in Theorem 5.1 is not a strict inclusion. In fact, it may be the case that a matrix M and its reduction $\mathscr{R}(M; S)$ have the same pseudospectra since the following example demonstrates.

Example 5.7. Consider the matrix $M \in \mathbb{C}^{4 \times 4}$ given by

$$M = \begin{bmatrix} 1 & 1 & 0 & 0 \\ 1 & 1 & 0 & 0 \\ 0 & 0 & 1 & 1 \\ 0 & 0 & 1 & 1 \end{bmatrix} \quad \text{and its reduction } \mathscr{R}(M; S) = \begin{bmatrix} \frac{\lambda}{\lambda-1} & 0 & 0 \\ 0 & 1 & 1 \\ 0 & 1 & 1 \end{bmatrix},$$

where $S = \{2, 3, 4\}$. Computing the 2-norm of the respective resolvents, we get

$$\|(M - \lambda I)^{-1}\| = \max(|\lambda|^{-1}, |\lambda - 2|^{-1}) \text{ and}$$
$$\|(\mathscr{R}(M; S) - \lambda I)^{-1}\| = \max(|\lambda|^{-1}, |\lambda - 2|^{-1}, |\lambda - 1||\lambda|^{-1}|\lambda - 2|^{-1}).$$

To show that the pseudospectra of M and $\mathscr{R}(M; S)$ are the same, we have only to demonstrate that the norms above are equal. We can do this by proving the inequality

$$|\lambda - 1||\lambda|^{-1}|\lambda - 2|^{-1} \leq \max(|\lambda|^{-1}, |\lambda - 2|^{-1}). \tag{5.7}$$

Notice that the triangle inequality implies

$$|\lambda - 1| \leq \frac{1}{2}|\lambda - 2| + \frac{1}{2}|\lambda| \leq \max(|\lambda|, |\lambda - 2|). \tag{5.8}$$

Inequality (5.7) follows for $\lambda \notin \{0, 2\}$ once we divide (5.8) by $|\lambda||\lambda - 2|$. Since $\{0, 2\} \subset \sigma(M), \sigma(\mathscr{R}(M;S))$, both 0 and 2 are included in the pseudospectra of these matrices. It then follows that $\sigma_\epsilon(M) = \sigma_\epsilon(\mathscr{R}(M;S))$ for all $\epsilon > 0$.

5.3 Inverse Pseudoeigenvalues

Recall that the inverse eigenvalues of a matrix $M(\lambda) \in \mathbb{W}_\pi^{n \times n}$ are the eigenvalues of its spectral inverse $\mathscr{S}^{-1}(M)$ (see Theorem 1.4 in Chap. 1). Thus, we may think of the "almost inverse eigenvalues" or inverse pseudoeigenvalues of $M(\lambda)$ as pseudoeigenvalues of $\mathscr{S}^{-1}(M)$. The precise definition is below, together with other equivalent definitions. These are analogous to Definitions 5.2(a)–(c) of pseudospectra.

Definition 5.3. Let $\epsilon > 0$. The set of ϵ–pseudoresonances of a matrix $M(\lambda) \in \mathbb{W}_\pi^{n \times n}$ is defined equivalently by the following conditions:

(a) *Resonance perturbation:*

$$\sigma_\epsilon^{-1}(M) = cl\big(\{\lambda \in \mathbb{C} : \|(M(\lambda) - \lambda I)^{-1}\mathbf{v}\| < \epsilon \text{ for some } \mathbf{v} \in \mathbb{C}^n \text{ with } \|\mathbf{v}\| = 1\}\big).$$

(b) *The inverse resolvent:*

$$\sigma_\epsilon^{-1}(M) = cl\big(\{\lambda \in \mathbb{C} : \|M(\lambda) - \lambda I\| > \epsilon^{-1}\}\big).$$

(c) *Perturbation of the spectral inverse:*

$$\sigma_\epsilon^{-1}(M) = cl\big(\{\lambda \in \mathbb{C} : \lambda \in \sigma(\mathscr{S}^{-1}(M) + E) \text{ for some } E \in \mathbb{C}^{n \times n} \text{ with } \|E\| < \epsilon\}\big).$$

We assume that the matrix and vector norms in (a)–(c) are compatible.

Note that Definition 5.3 is simply Definition 5.2 in which $M(\lambda)$ is replaced by the matrix $\mathscr{S}^{-1}(M)$ on the right-hand side of parts (a)–(c). Hence, the equivalence of Definitions 5.3(a)–(c) follow from arguments similar to those in 5.4. Moreover, since $\sigma^{-1}(M) = \sigma(\mathscr{S}^{-1}(M))$ for every $M(\lambda) \in \mathbb{W}_\pi^{n \times n}$, then using the matrix $E \equiv 0$ in Definition 5.3(c) yields

$$\sigma^{-1}(M) \subset \sigma_\epsilon^{-1}(M) \text{ for each } \epsilon > 0.$$

Additionally, observe that if $w(\lambda) = p(\lambda)/q(\lambda) \in \mathbb{W}_\pi$, then by definition, $\pi(p) \leq \pi(q)$. Hence we have the limit

Fig. 5.5 Pseudoresonance regions of the matrix $\mathscr{R}(M;S)$ given in Example 5.3 for $\epsilon = 1/3$ (*blue*), $\epsilon = 1/4$ (*red*), and $\epsilon = 1/5$ (*tan*), using the matrix 1-norm. The eigenvalues $\sigma(M_{\bar{S}\bar{S}}) = \{0, 1\}$ are indicated

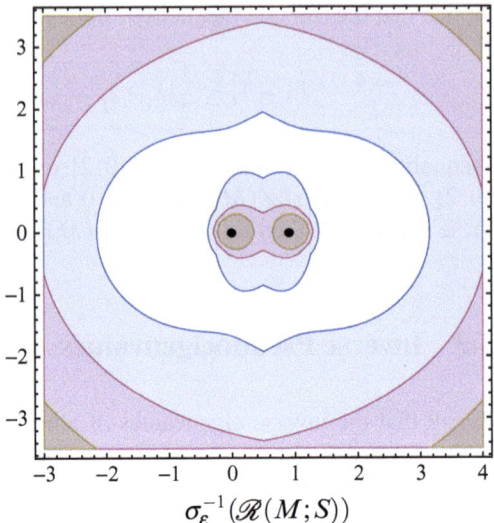

$$\lim_{|\lambda|\to\infty} |w(\lambda)| = c,$$

for some constant $c \geq 0$. Therefore, for each $M(\lambda) \in \mathbb{W}_\pi^{n\times n}$, $||M(\lambda) - \lambda I|| = \mathscr{O}(\lambda)$ for large λ. This leads to the following remark.

Remark 5.2. If $M \in \mathbb{W}_\pi^{n\times n}$, then the value $\lambda = \infty$ is always an inverse pseudoeigenvalue of M. This means that for each $\epsilon > 0$, the set $\sigma_\epsilon^{-1}(M)$ contains the complement of a ball centered at the origin with sufficiently large radius. (See Fig. 5.5, for example.)

Example 5.8. In Fig. 5.5, we show the pseudoresonance regions of the matrix $\mathscr{R}(M;S)$ from Example 5.3 for $\epsilon = 1, 1/2, 1/4$. As can be computed, the inverse spectrum of $\mathscr{R}(M;S)$ is empty. However, the inverse ϵ-pseudoeigenvalue regions reveal that the eigenvalues $\sigma(M_{\bar{S}\bar{S}}) = \{0, 1\}$ act like inverse eigenvalues. Specifically, $\sigma(M_{\bar{S}\bar{S}}) \subset \sigma_\epsilon^{-1}(\mathscr{R}(M;S))$ for each ϵ that we consider.

As it turns out, the situation in Example 5.8 does not hold for every matrix reduction. Similar to Example 5.5, if

$$M = \begin{bmatrix} 1 & 1 \\ 0 & 0 \end{bmatrix},$$

and we consider the sets $S = \{1\}$ and $\bar{S} = \{2\}$, then one can show that $\sigma(M_{\bar{S}\bar{S}}) = \{0\}$ is not contained in $\sigma_\epsilon^{-1}(\mathscr{R}(M;S))$ for small $\epsilon > 0$. That is, the eigenvalues $\sigma(M_{\bar{S}\bar{S}})$ do not always act like inverse eigenvalues of the matrix $\mathscr{R}(M;S)$.

As with the pseudospectra studied in Sect. 5.1 of this chapter, we give a physical interpretation of inverse pseudospectra using a mass–spring system.

5.3 Inverse Pseudoeigenvalues

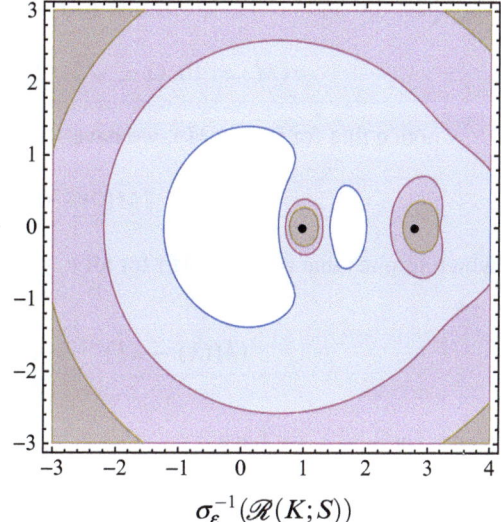

Fig. 5.6 Pseudoresonance regions of the matrix $\mathscr{R}_\lambda(K;S)$ given in Example 5.4, with $S = \{1,4\}$ for $\epsilon = 1/2$ (*blue*), $\epsilon = 1/3$ (*red*), and $\epsilon = 1/4$ (*tan*). The resonances $\sigma^{-1}(\mathscr{R}_\lambda(K;S)) = \{1,3\}$ are indicated

$\sigma_\epsilon^{-1}(\mathscr{R}(K;S))$

Example 5.9. The mass–spring system considered in Example 5.4 has inverse eigenvalues when restricted to the set of boundary nodes $S \subset \{1,4\}$. The inverse pseudoeigenvalues of the reduced system correspond to frequencies for which there is a displacement on the boundary that generates relatively "large" forces at these nodes. For this reason, we may think of inverse eigenvalues as *resonances* of the system, and inverse pseudoeigenvalues as *pseudoresonances*. In Fig. 5.6, we display some pseudoresonance regions of the mass–spring system restricted to the set $S = \{1,4\}$.

Since we allow ϵ to be any positive value, there is nothing preventing an eigenvalue of a matrix M from also being an ϵ-pseudoresonance of M (or a resonance from being a ϵ-pseudoeigenvalue). In other words, we could have an $\epsilon > 0$ for which

$$\sigma^{-1}(M) \cap \sigma_\epsilon(M) \neq \emptyset \text{ or } \sigma(M) \cap \sigma_\epsilon^{-1}(M) \neq \emptyset,$$

as the following example shows.

Example 5.10. Consider the matrix $M(\lambda) \in \mathbb{W}_\pi^{2\times 2}$ given by

$$M(\lambda) = \begin{bmatrix} \frac{1}{\lambda-1} & 0 \\ 0 & 0 \end{bmatrix}.$$

The spectrum and inverse spectrum of $M(\lambda)$ are respectively

$$\sigma(M) = \{0, (1 \pm \sqrt{5})/2\} \text{ and } \sigma^{-1}(M) = \{1\}.$$

Now notice that for $0 \in \sigma(M)$, we have

$$\|M(0) - 0I\| = 1,$$

which implies that $0 \in \sigma_\epsilon^{-1}(M)$ for all $\epsilon \geq 1$. The resolvent of M is

$$(M(\lambda) - \lambda I)^{-1} = \begin{bmatrix} \frac{\lambda-1}{-\lambda^2+\lambda+1} & 0 \\ 0 & -\frac{1}{\lambda} \end{bmatrix}.$$

Hence for $\lambda = 1$, we have

$$\|(M(1) - I)^{-1}\| = 1,$$

which means that $1 \in \sigma_\epsilon(A)$ for all $\epsilon \geq 1$.

Because the inverse pseudoeigenvalues of a matrix $M \in \mathbb{W}_\pi^{n \times n}$ can be defined in terms of the pseudoeigenvalues of the spectral inverse $\mathscr{S}^{-1}(M)$, we can generalize Theorem 1.4 as follows.

Theorem 5.2. *Suppose $M(\lambda) \in \mathbb{W}_\pi^{n \times n}$ and $\epsilon > 0$. Then*

$$\sigma_\epsilon^{-1}(M) = \sigma_\epsilon(\mathscr{S}^{-1}(M)) \text{ and } \sigma_\epsilon(M) = \sigma_\epsilon^{-1}(\mathscr{S}^{-1}(M)).$$

Proof. Let $M(\lambda) \in \mathbb{W}_\pi^{n \times n}$ and $\epsilon > 0$. Observe that

$$\sigma_\epsilon(M) = cl\big(\{\lambda \in \mathbb{C} : \|(M(\lambda) - \lambda I)^{-1}\| > \epsilon^{-1}\}\big); \text{ and}$$

$$\sigma_\epsilon^{-1}(\mathscr{S}^{-1}(M)) = cl\big(\{\lambda \in \mathbb{C} : \|\mathscr{S}^{-1}(M) - \lambda I\| > \epsilon^{-1}\}\big),$$

using Definitions 5.2(b) and 5.3(b), respectively. Since $\mathscr{S}^{-1}(M) - \lambda I = (M(\lambda) - \lambda I)^{-1}$, then $\sigma_\epsilon(M) = \sigma_\epsilon^{-1}(S(M))$. The equality $\sigma_\epsilon^{-1}(M) = \sigma_\epsilon(\mathscr{S}^{-1}(M))$ similarly follows. □

Because of the seemingly invertible relationship between pseudospectra and inverse pseudospectra in Theorem 5.2, it is tempting to think of the ϵ−pseudoresonances of a matrix as the complement of its $1/\epsilon$−pseudoeigenvalues. However, the two are not always equal, as can be seen in the following theorem.

Theorem 5.3. *For $M(\lambda) \in \mathbb{W}_\pi^{n \times n}$, let $\epsilon > 0$. Then $cl\big(\overline{\sigma_{1/\epsilon}(M)}\big) \subseteq \sigma_\epsilon^{-1}(M)$. However, the reverse inclusion does not hold in general.*

5.3 Inverse Pseudoeigenvalues

This theorem means that in general, there is not enough information in the pseudispectra of a matrix to reconstruct its pseudoresonances. We now proceed with the proof of Theorem 5.3.

Proof. For $M(\lambda) \in \mathbb{W}_\pi^{n \times n}$ and a matrix norm $||\cdot||$, the inequality

$$||M(\lambda) - \lambda I||^{-1} \leq ||(M(\lambda) - \lambda I)^{-1}|| \tag{5.9}$$

holds for every $\lambda \in \text{dom}(M) - \sigma(M)$. Let int($\Omega$) denote the *interior* of the set $\Omega \subseteq \mathbb{C}$, i.e., the largest open subset of Ω. For $\epsilon > 0$, using Definition 5.2(b), we obtain

$$cl\left(\overline{\sigma_{1/\epsilon}(M)}\right) = cl\left(\overline{cl(\{\lambda \in \mathbb{C} : ||(M(\lambda) - \lambda I)^{-1}|| > \epsilon\})}\right)$$
$$= cl\left(\text{int}(\{\lambda \in \mathbb{C} : ||(M(\lambda) - \lambda I)^{-1}|| \leq \epsilon\})\right)$$
$$= cl\left(\{\lambda \in \mathbb{C} : ||(M(\lambda) - \lambda I)^{-1}|| \leq \epsilon\}\right).$$

Similarly, it follows from Definition 5.3(b) that

$$\sigma_\epsilon^{-1}(M) = cl\left(\{\lambda \in \mathbb{C} : ||M(\lambda) - \lambda I|| > \epsilon^{-1}\}\right)$$
$$= cl\left(\{\lambda \in \mathbb{C} : ||M(\lambda) - \lambda I||^{-1} \leq \epsilon\}\right).$$

By inequality (5.9), we have

$$\{\lambda \in \mathbb{C} : ||(M(\lambda) - \lambda I)||^{-1} \leq \epsilon\} \subseteq \{\lambda \in \mathbb{C} : ||(M(\lambda) - \lambda I)^{-1}|| \leq \epsilon\}$$

implying the first half of the result.

To show that the reverse inclusion does not hold in general, take, for instance, the matrix $M(\lambda)$ from Example 5.10. It is easy to compute $||M(2) - 2I|| = 2$ and $||(M(2) - 2I)^{-1}|| = 1$. Taking $\epsilon = 2/3$, we have $2 \in \sigma_{2/3}^{-1}(M) \cap \sigma_{3/2}(M)$. □

Before finishing this section, we note that Theorem 5.1 states that the ϵ-pseudospectrum of a matrix becomes a subset of this region, since the matrix is reduced. However, for ϵ-pseudoresonances of a matrix, there is no such inclusion result, as is demonstrated in the following example.

Example 5.11. Consider the matrix $M \in \mathbb{C}^{4 \times 4}$ and its reduction $\mathscr{R}(M; S)$ given by

$$M = \begin{bmatrix} 0 & 1 & 1 & 0 \\ 1 & 0 & 0 & 0 \\ 1 & 0 & 0 & 1 \\ 0 & 0 & 1 & 0 \end{bmatrix} \quad \text{and} \quad \mathscr{R}(M; S) = \begin{bmatrix} \frac{1}{\lambda} & 1 \\ 1 & \frac{1}{\lambda} \end{bmatrix},$$

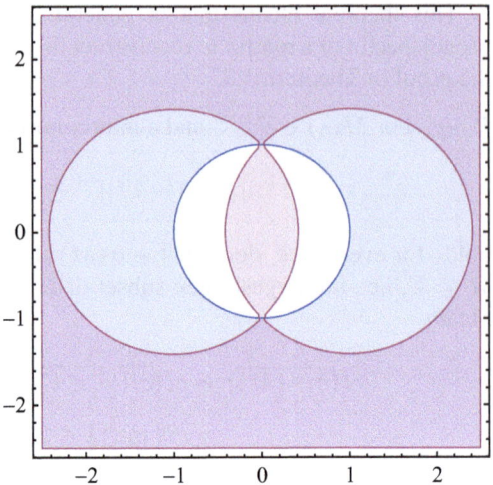

Fig. 5.7 The pseudoresonance regions $\sigma_\epsilon^{-1}(M)$ (*blue*) and $\sigma_\epsilon^{-1}(\mathscr{R}(M;S))$ (*red*) are shown for $\epsilon = 1/3$, where M and $\mathscr{R}(M;S)$ are the matrices considered in example 5.11

where $S = \{1, 4\}$. The ϵ-pseudoresonances $\sigma_\epsilon^{-1}(M)$ and $\sigma_\epsilon^{-1}(\mathscr{R}(M;S))$ are shown in Fig. 5.7 for $\epsilon = 1/3$ in blue and red, respectively. As can be seen,

$$\sigma_\epsilon^{-1}(M) \not\subset \sigma_\epsilon^{-1}(\mathscr{R}(M;S)) \text{ and } \sigma_\epsilon^{-1}(\mathscr{R}(M;S)) \not\subset \sigma_\epsilon^{-1}(M).$$

Hence, the inclusion result for pseudospectra given in Theorem 5.1 does not hold for pseudoresonances.

5.4 Eigenvalue Inclusions and Equivalence of Definitions

In this section, we show that the three pseudoeigenvalue regions given in Definition 5.2(a)–(c) coincide and include the eigenvalues of the matrix. To do so, we define the following regions. For $M(\lambda) \in \mathbb{W}_\pi^{n \times n}$ and $\epsilon > 0$, let

$$\sigma_{\epsilon,a}(M) = \{\lambda \in \mathbb{C} : ||(M(\lambda) - \lambda I)\mathbf{v}|| < \epsilon \text{ for some } \mathbf{v} \in \mathbb{C}^n \text{ with } ||\mathbf{v}|| = 1\}$$
(5.10)

$$\sigma_{\epsilon,b}(M) = \{\lambda \in \mathbb{C} : ||(M(\lambda) - \lambda I)^{-1}|| > \epsilon^{-1}\} \cup \sigma(M) \tag{5.11}$$

$$\sigma_{\epsilon,c}(M) = \{\lambda \in \mathbb{C} : \lambda \in \sigma(M(\lambda) + E) \text{ for some } E \in \mathbb{C}^{n \times n} \text{ with } ||E|| < \epsilon\}.$$
(5.12)

Note that the regions $\sigma_{\epsilon,a}(M)$, $\sigma_{\epsilon,b}(M)$, and $\sigma_{\epsilon,c}(M)$ are the regions given in Definition 5.2(a)–(c) without taking the closure. Before proving the main result of this section, we require the following result.

5.4 Eigenvalue Inclusions and Equivalence of Definitions

Lemma 5.1. *Let $\epsilon > 0$. If $M(\lambda) = M$ is a complex-valued matrix, then the regions $\sigma_{\epsilon,a}(M)$, $\sigma_{\epsilon,b}(M)$, and $\sigma_{\epsilon,c}(M)$ coincide.*

The reason lemma 5.1 holds if $M \in \mathbb{C}^{n \times n}$ is that in this case, $\sigma_{\epsilon,a}(M)$, $\sigma_{\epsilon,b}(M)$, and $\sigma_{\epsilon,c}(M)$ are the same as the set(s) $\sigma_\epsilon(M)$ given in Definition 5.1 in (a), (b), and (c), respectively. With this in place, we are now in a position to prove the following theorem.

Theorem 5.4. *Let $M(\lambda) \in \mathbb{W}_\pi^{n \times n}$ and $\epsilon > 0$. Then the regions given in Definition 5.2(a)–(c) are equivalent. Moreover, $\sigma(M) \subset \sigma_\epsilon(M)$.*

Proof. Suppose $M(\lambda) \in \mathbb{W}_\pi^{n \times n}$ and $\epsilon > 0$. If $\lambda_0 \in \sigma(M)$, then there is a unit vector $\mathbf{v} \in \mathbb{C}^n$ such that

$$(M(\lambda) - \lambda I)\mathbf{v} = w(\lambda) \in \mathbb{W}_\pi^n,$$

where $w(\lambda_0) = 0$. Since $\sigma(M)$ and $\overline{\text{dom}(M)}$ are finite, there is a neighborhood $U \ni \lambda_0$ such that for $\tilde{U} = U - \{\lambda_0\}$, the following hold:

(i) $\tilde{U} \subset \text{dom}(M)$;
(ii) $\|w(\lambda)\| < \epsilon$ for $\lambda \in \tilde{U}$; and
(iii) $(\sigma(M) - \{\lambda_0\}) \cap \tilde{U} = \emptyset$.

In particular, (ii) implies the set inclusion $\tilde{U} \subset \sigma_{\epsilon,a}(M)$.

For each $\lambda \in \text{dom}(M)$, observe that the matrix $M(\lambda)$ is in $\mathbb{C}^{n \times n}$. Since the regions given in (5.10)–(5.12) are equivalent for every complex-valued matrix, Lemma 5.1 implies

$$\tilde{U} \subset \sigma_{\epsilon,a}(M) - \{\lambda_0\}, \sigma_{\epsilon,b}(M) - \{\lambda_0\}, \sigma_{\epsilon,c}(M) - \{\lambda_0\}.$$

This, in turn, implies

$$\sigma(M) \subset cl\left(\sigma_{\epsilon,a}(M)\right), cl\left(\sigma_{\epsilon,b}(M) - \sigma(M)\right), cl\left(\sigma_{\epsilon,c}(M)\right). \qquad (5.13)$$

In particular, if $\sigma_{\epsilon,b}(M) - \sigma(M)$ is open, then $\sigma_{\epsilon,b}(M)$ is open.

Note that the norm of a vector or matrix is continuous with respect to its entries. Also, the eigenvalues of a matrix depend continuously on the matrix entries. Thus, the sets $\sigma_{\epsilon,a}(M)$, $\sigma_{\epsilon,b}(M) - \sigma(M)$, and $\sigma_{\epsilon,c}(M)$ are open. Therefore, the set $\sigma_{\epsilon,b}(M)$ is also open.

Since the sets given in (5.10)–(5.12) are equivalent on $\text{dom}(M)$, and $\overline{\text{dom}(M)}$ is a finite set, it follows that

$$\sigma_{\epsilon,a}(M) \cap \text{dom}(M) = \sigma_{\epsilon,b}(M) \cap \text{dom}(M) = \sigma_{\epsilon,c}(M) \cap \text{dom}(M)$$

is an open set. Taking the closure, it follows that

$$cl\big(\sigma_{\epsilon,a}(M)\big) = cl\big(\sigma_{\epsilon,b}(M) - \sigma(M)\big) = cl\big(\sigma_{\epsilon,c}(M)\big),$$

implying that Definitions 5.2(a)–(c) are equivalent. Moreover, equation (5.13) implies $\sigma(M) \subset \sigma_{\epsilon}(M)$. This completes the proof. □

The proof that Definitions 5.3(a)–(c) are equivalent is very similar to the proof of Theorem 5.4 and is therefore omitted.

Chapter 6
Improved Estimates of Survival Probabilities

In the theory of dynamical systems, the main object of study is the orbits of the elements of a set under a fixed rule. Most often, this means that for a given function $f : M \to M$, we would like to understand the behavior of the sequence $\{f^k(x) : k \geq 0\}$ for each $x \in M$. However, not all systems have this form.

If $f : \tilde{M} \to M$, where $\tilde{M} \subset M$, then it is possible that a point does not have an orbit under this rule. That is, it may happen that at some moment in time $k < \infty$, the iterate $f^k(x)$ is in $M - \tilde{M}$, so that $f^{k+1}(x)$ is undefined. This is the situation in the theory of open dynamical systems.

In this theory, the map $f : \tilde{M} \to M$ induces an *open dynamical system* in which the point $x \in \tilde{M}$ is said to *escape* at time k if $f^k(x) \in M - \tilde{M}$. Here, the set $M - \tilde{M}$ is referred to as a *hole*, since it is the set through which the point escapes the system.

If the point $x \in \tilde{M}$ escapes the system, then we no longer consider what happens to it beyond that point in time. On the other hand, if $f^k(x) \in \tilde{M}$, then we say that the point x has *survived* up to time k. If the point $x \in \tilde{M}$ survives for all $k > 0$, then we say that it survives for all time.

For an open dynamical system $f : \tilde{M} \to M$, one of the most basic questions that can be asked is this: what is the probability of a point surviving in the system for all time? More precisely, if μ is some probability measure on M, then what is the measure of the set of points that never escape the system? A natural extension of this question is, what is the probability of a point surviving in the system until some time $k < \infty$?

The first question addresses the *asymptotic chances* of surviving, while the second considers the *finite-time probability* that a typical point will remain in the system until some time $k < \infty$. The reason we consider these questions is that it is possible to associate the dynamics of an open dynamical system with the dynamics of a related dynamical network. This relationship between an open dynamical system and a dynamical network will allow us to estimate, and in some cases give precisely, the asymptotic and finite-time survival probabilities of a general class of open systems.

Importantly, because of the interplay between the theory of open dynamical systems and dynamical networks, it is possible to use techniques similar to those introduced in Chap. 3 to analyze these survival probabilities. Our main result in this direction is that there is a number of transformations that can be used to sharper survival probability estimates of open systems we consider here.

The type of systems we consider is that of one-dimensional maps that are piecewise smooth, and have a finite Markov partition. The holes in these systems consist of a collection of these partition elements and therefore have a size that is both finite and fixed.

The reason we consider only one-dimensional systems in this chapter is simply for the sake of clarity. The results that we give here can be extended, with only slight modifications, to higher-dimensional systems.

6.1 Open Dynamical Systems

In this section, we define the basic concepts that we will use throughout the chapter. We begin by introducing the type of systems we will use to generate the open dynamical systems that we will consider.

Let $f : I \to I$, where $I = [0, 1]$. For $0 = q_0 < q_1 < \cdots < q_{m-1} < q_m = 1$, we let $\xi_i = (q_{i-1}, q_i]$ for each $i \in M = \{1, \ldots, m\}$ and assume that the following hold. First, the function $f|_{\xi_i}$ is differentiable for each $1 \leq i \leq m$. Second, the sets $\xi_i = (q_{i-1}, q_i]$ form a *Markov partition* $\xi = \{\xi_i\}_{i=1}^m$ of f. That is, for each $i \in M$, the closure $cl(f(\xi_i))$ is the interval $[q_j, q_{j+k}]$ for some $k \geq 1$ and j that depends on i.

Given this setup, we consider the situation in which orbits of $f : I \to I$ escape through an element of the Markov partition $\xi = \{\xi_i\}_{i=1}^m$ or, more generally, some union H of these partition elements. Equivalently, we can modify the function f so that orbits cannot leave the set H once they have entered it. With this approach, orbits that enter H are considered to have escaped from the system. This latter approach, of modifying $f|_H$, turns out to be more convenient for our discussion and will be used to formally define the open systems that we consider.

Definition 6.1. Let $f : I \to I$ have the Markov partition $\xi = \{\xi_i\}_{i=1}^m$. If $H = \bigcup_{i \in \mathscr{I}} \xi_i$, where $\mathscr{I} \subset M$, then we let $f_H : I \to I$ be the map defined by

$$f_H(x) = \begin{cases} f(x) & \text{if } x \notin H, \\ x & \text{otherwise.} \end{cases}$$

We call the set H a *hole* and the function $f_H : I \to I$ the *open dynamical system* (f_H, I) generated by the (*closed*) dynamical system (f, I) over H.

6.2 Piecewise Linear Functions

Note that the partition ξ is still a Markov partition of the open dynamical system (f_H, I), but the dynamics of the original system (f, I) have been modified, so that each point in H is now a fixed point of the open system (f_H, I).

For $n \geq 0$, let

$$\mathbb{E}^n(f_H) = \{x \in I : f^n(x) \in H, \ f^k(x) \notin H, \ 0 \leq k < n\}$$
$$= \{x \in I : f_H^n(x) \in H, \ f_H^k(x) \notin H, \ 0 \leq k < n\}; \text{ and}$$
$$\mathbb{T}^n(f_H) = \{x \in I : f^k(x) \in H, \text{ for some } k, \ 0 \leq k \leq n\}$$
$$= \{x \in I : f_H^k(x) \in H, \text{ for some } k, \ 0 \leq k \leq n\}.$$

The set $\mathbb{E}^n(f_H)$ consists of those points that escape through the hole H at time n, while $\mathbb{T}^n(f_H)$ comprises those points that escape through H before time $n+1$.

For the moment, suppose μ is a probability measure on I, i.e., $\mu(I) = 1$. Then $\mu(\mathbb{E}^n(f_H))$ can be treated as the probability that an orbit of f enters H for the first time at time n, and $\mu(\mathbb{T}^n(f_H))$ the probability that an orbit of f enters H before time $n+1$. In this regard,

$$\mathbb{P}^n(f_H) = 1 - \mu(\mathbb{T}^n(f_H))$$

represents the probability that a typical point of I does not fall into the hole H by time n. For this reason, the quantity $\mathbb{P}^n(f_H)$ is called the *survival probability*, at time n, of the dynamical system f_H for the measure μ. Thus, the probability

$$\mathbb{P}(f_H) = \lim_{n \to \infty} \mathbb{P}^n(f_H)$$

is the probability that a point survives in the open system (f_H, I) for all time.

One of the fundamental problems in the theory of open systems is finding ways of approximating $\mu(\mathbb{E}^n(f_H))$ and $\mu(\mathbb{T}^n(f_H))$ for a finite $n \geq 0$. In the following section, we give exact formulas for these quantities in the case that f_H is a piecewise linear function with nonzero slope and μ is Lebesgue measure. In Sect. 6.2, we remove the assumption that f_H is piecewise linear and present a method for estimating $\mu(\mathbb{E}^n(f_H))$ and $\mu(\mathbb{T}^n(f_H))$ for functions that are nonlinear.

6.2 Piecewise Linear Functions

As a first step in investigating the survival probabilities of an open dynamical system, we consider those systems (f_H, I) that are linear when restricted to the elements of the partition ξ. More formally, suppose $H = \bigcup_{i \in \mathscr{I}} \xi_i$ for some $\mathscr{I} \subset M$. Let $\mathscr{L}$ be the set of all open systems (f_H, I) such that

$$|f_H'(x)| = c_i > 0 \text{ for } x \in \xi_i \text{ and } i \notin \mathscr{I},$$

where each c_i is in $\mathbb{R}$. The set $\mathscr{L}$ is then the collection of all open systems that have a nonzero constant slope when restricted to any $\xi_i \not\subseteq H$.

To each $f_H \in \mathscr{L}$ there is an associated matrix, which can be used to compute both $\mu(\mathbb{E}^n(f_H))$ and $\mu(\mathbb{T}^n(f_H))$. To define this matrix, let

$$\xi_{ij} = \xi_i \cap f^{-1}(\xi_j) \text{ for } 1 \le i, j \le m. \tag{6.1}$$

Definition 6.2. Suppose $f_H \in \mathscr{L}$, where $H = \bigcup_{i \in \mathscr{I}} \xi_i$ for some $\mathscr{I} \subset M$. The matrix $A_H \in \mathbb{R}^{m \times m}$ given by

$$(A_H)_{ij} = \begin{cases} |f'(x)|^{-1} & \text{for } x \in \xi_{ij} \ne \emptyset, \ i \notin \mathscr{I}, \\ 0 & \text{otherwise}, \end{cases} \quad 1 \le i, j \le m,$$

is called the *weighted transition matrix* of f_H.

Associated with the open system (f_H, I) there is also an unweighted directed graph $\Gamma = (V, E_H)$ with *vertices* V and *edges* E_H. As before, if $V = \{v_1, \ldots, v_m\}$, then we let e_{ij} denote the edge from vertex v_i to v_j.

Definition 6.3. Let $f_H : I \to I$, where $H = \bigcup_{i \in \mathscr{I}} \xi_i$ for some $\mathscr{I} \subset M$. We define $\Gamma_H = (V, E_H)$ to be the graph with

(a) vertices $V = \{v_1, \ldots, v_m\}$,
(b) edges $E_H = \{e_{ij} : cl(\xi_j) \subseteq cl(f(\xi_i)), \ i \notin \mathscr{I}\}$.

The graph Γ_H is called the *transition graph* of f_H.

The vertex set $V = \{v_1, \ldots, v_m\}$ of Γ_H represents the elements of the Markov partition $\xi = \{\xi_i\}_{i=1}^m$, and the edge set E_H, the possible transitions between elements of ξ. Hence $e_{ij} \in E_H$ only if there is an $x \in \xi_i \not\subseteq H$ such that $f_H(x) \in \xi_j$, i.e., if it is possible to transition from ξ_i to ξ_j. We note that since $H = \emptyset$ is a possible hole, the original (closed) system (f, I) has a well-defined transition graph, which we denote by Γ.

Note that the graph Γ_H does not carry the same amount of information as the matrix A_H. The reason is that Γ_H designates only how orbits can transition between elements of ξ, whereas A_H additionally gives each of these transitions a weight. That is, the adjacency matrix $M(\Gamma_H)$ is not equal to A_H.

However, the graph Γ_H does give us a way of visualizing how orbits escape from the system. This will be useful in Sect. 6.4, where we use the system's graph structure to improve our estimate of the system's survival probabilities.

The transition graph of an open system is given in the following example.

Example 6.1. Let the function $f : I \to I$ be the tent map

$$f(x) = \begin{cases} 2x, & 0 \le x \le 1/2, \\ 2 - 2x, & 1/2 < x \le 1, \end{cases}$$

6.2 Piecewise Linear Functions

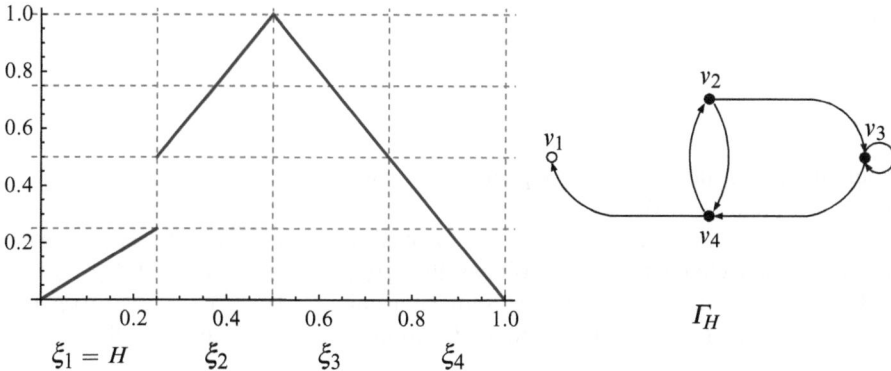

Fig. 6.1 The the open system (f_H, I) (*left*) and its transition graph Γ_H (*right*) considered in Example 6.3

with Markov partition $\xi = \{(0, 1/4], (1/4, 1/2], (1/2, 3/4], (3/4, 1]\}$. Here, we let H be the hole $H = (0, 1/4]$. We then have the open system $f_H : I \to I$ shown in Fig. 6.3 (left) with the graph of transitions Γ_H (right).

We emphasize the fact that $H = \xi_1$ is the hole in the open system f_H by drawing the vertex v_1 as an open circle in Γ_H. Note that the only difference between the transition graph Γ of $f : I \to I$ and Γ_H is that there are no edges originating from v_1 in Γ_H. In this sense, a hole H is an absorbing state, since nothing leaves H once it enters (Fig. 6.1).

We now consider how to compute the quantities $\mathbb{E}^n(f_H)$ and $\mathbb{T}^n(f_H)$ for an open system $f_H \in \mathscr{L}$. To do this, we let $\mathbf{1} = [1, \ldots, 1]$ be the $1 \times m$ vector of ones, and $\mathbf{e}_H$ the $m \times 1$ vector given by

$$(\mathbf{e}_H)_i = \begin{cases} \mu(\xi_i) & \text{if } i \in \mathscr{I}, \\ 0 & \text{otherwise.} \end{cases}$$

For the sake of simplicity, we let μ denote Lebesgue measure for the remainder of this chapter. With this in place, we have the following theorem.

Theorem 6.1. *If $f_H \in \mathscr{L}$ and $n \geq 0$, then*

$$\mu(\mathbb{E}^n(f_H)) = \mathbf{1} A_H^n \mathbf{e}_H;$$

$$\mu(\mathbb{T}^n(f_H)) = \mathbf{1} \left(\sum_{i=0}^{n} A_H^i \right) \mathbf{e}_H.$$

Proof. For the open system $f_H : I \to I$ with Markov partition $\xi = \cup_{i=1}^{m} \xi_i$ and hole $H = \cup_{i \in \mathscr{I}} \xi$ for $\mathscr{I} \subseteq M$, let

$$\xi_{i_0 i_1 \ldots i_n} = \{x \in I : f^k(x) \in \xi_{i_k} \text{ for } 0 \leq k \leq n\}.$$

152 6 Improved Estimates of Survival Probabilities

Then for $n > 0$, the set $\mathbb{E}^n(f_H)$ has the measure given by

$$\mu(\mathbb{E}^n(f_H)) = \sum_{\gamma} \mu(\xi_\gamma), \tag{6.2}$$

where the sum is taken over all sequences $\gamma = i_0 i_1 \ldots i_n$ with $\xi_{i_k} \not\subset H$ for $0 \le k < n$, $\xi_{i_n} \subset H$, and $\xi_\gamma \ne \emptyset$.

Let $z_i = |f'(x)|^{-1}$, where $x \in \xi_i$ for each $i \in M$. Since f_H is piecewise linear on each element of ξ, it follows that $\mu(\xi_{i_0 i_1}) = z_{i_0} \mu(\xi_{i_1})$ if $\xi_{i_0 i_1} \ne \emptyset$, and $\mu(\xi_{i_0 i_1}) = 0$ otherwise. By the same reasoning, $\mu(\xi_{i_0 i_1 i_2}) = z_{i_0} z_{i_1} \mu(\xi_{i_2})$ if $\xi_{i_0 i_1 i_2} \ne \emptyset$, and $\mu(\xi_{i_0 i_1 i_2}) = 0$ otherwise. Continuing in this manner, it follows that

$$\mu(\xi_{i_0 i_1 \ldots i_n}) = \begin{cases} \left(\prod_{k=0}^{n-1} z_{i_k}\right) \mu(\xi_{i_n}) & \text{if } \xi_{i_0 i_1 \ldots i_n} \ne \emptyset, \\ 0 & \text{otherwise.} \end{cases}$$

Combining this with equation (6.2), we have

$$\mu(\mathbb{E}^n(f_H)) = \sum_{\ell=1}^{m} \left(\sum_{\gamma_\ell} \left(\prod_{k=0}^{n-1} z_{i_k} \right) \mu(\xi_{i_n}) \right), \tag{6.3}$$

where the second sum is taken over all $\gamma_\ell = i_0 i_1 \ldots i_n$ with $i_0 = \ell$, $\xi_{i_k} \not\subset H$ for all $0 \le k < n$, $\xi_{\gamma_\ell} \ne \emptyset$, and $\xi_{i_n} \subset H$.

For $1 \le \ell \le m$, let $(A_H^n \mathbf{e}_H)_\ell$ denote the ℓth component of $A_H^n \mathbf{e}_H \in \mathbb{R}^m$. From the definition of matrix multiplication, it follows that

$$(A_H^n \mathbf{e}_H)_\ell = \sum_{\gamma_\ell} \left(\prod_{k=0}^{n-1} (A_H)_{i_k i_{k+1}} \right) (\mathbf{e}_H)_{i_n}, \tag{6.4}$$

where the sum is taken over all $\gamma_\ell = i_0 i_1 \ldots i_n$ with $i_0 = \ell$.

Note that $(A_H)_{i_k i_{k+1}} = 0$ if $\xi_{i_k} \subset H$ or $\xi_{i_k i_{k+1}} = \emptyset$. Also, $(\mathbf{e}_H)_{i_n} = \mu(\xi_{i_n}) = 0$ if $\xi_{i_n} \not\subset H$. Hence, the sum in equation (6.4) can be taken over all $\gamma_\ell = i_0 i_1 \ldots i_n$ in which $i_0 = \ell$, $\xi_{i_k} \not\subset H$ for $0 \le k < n$, $\xi_{\gamma_\ell} \ne \emptyset$, and $\xi_{i_n} \subseteq H$. Since $(A_H)_{i_k i_{k+1}} = z_{i_k}$ for each $0 \le k < n$ under the assumption that $\xi_{\gamma_\ell} \ne \emptyset$ then, from (6.3) and (6.4), it follows that

$$\mu(\mathbb{E}^n(f_H)) = \sum_{\ell=1}^{m} \left(\sum_{\gamma_\ell} \left(\prod_{k=0}^{n-1} (A_H)_{i_k i_{k+1}} \right) \mu(\xi_{i_n}) \right)$$

$$= \sum_{\ell=1}^{m} (A_H^n \mathbf{e}_H)_\ell$$

$$= \mathbf{1} A_H^n \mathbf{e}_H.$$

6.2 Piecewise Linear Functions

For the case $n = 0$, we have that $\mu(\mathbb{E}^0(f_H)) = \mu(H)$. Since

$$\mathbf{1}A_H^0 \mathbf{e}_H = \mathbf{1}\mathbf{e}_H = \mu(H),$$

then $\mu(\mathbb{E}^n(f_H)) = \mathbf{1}A_H^n \mathbf{e}_H$ for all $n \geq 0$. Using this equality, we have

$$\mu(\mathbb{T}^n(f_H)) = \sum_{i=0}^{n} \mu(\mathbb{E}^i(f_H)) = \sum_{i=0}^{n}(\mathbf{1}A_H^n \mathbf{e}_H) = \mathbf{1}\sum_{i=0}^{n}(A_H^n)\mathbf{e}_H$$

by the linearity of matrix multiplication. This verifies the result. □

Before moving on, we mention that in the theory of open dynamical systems, the function $f_H : I \to I$ is typically assumed to have the property $\mathbb{P}(f_H) = 0$ for a given hole H, so that almost every point escapes the system as $n \to \infty$. This is done, for instance, by assuming that f is an expanding map with $|f'| > 1$.

This is assumed because otherwise, there is a set $I_* \subset I$ of positive measure that never escapes through H. If this were the case, then rather than investigating the system (f_H, I), we could consider the open system $(\tilde{f}_H, \tilde{I})$, where $\tilde{I} = I - I_*$, in which $\mathbb{P}(\tilde{f}_H) = 0$.

However, a nice feature of Theorem 6.1 is that it can be used whether or not $\mathbb{P}(f_H) = 0$. In fact, if it is the case that $\mathbb{P}(f_H) = p < 1$, then we can simply consider the new measure $\nu = \mu/p$ on the system f_H, so that using this scaled version of Lebesgue measure, we have

$$\mathbb{P}_\nu(f_H) = 1 - \nu(\mathbb{T}^n(f_H)) = 0.$$

In this sense, Theorem 6.1 gives us two pieces of useful information: the Lebesgue measure of those points that escape through H at any moment in time and ultimately the Lebesgue measure of those points that survive for all time.

Recall that the eigenvalues and spectral radius of a matrix $M \in \mathbb{C}^{n \times n}$ are denoted by $\sigma(M)$ and $\rho(M)$, respectively. The following are corollaries to Theorem 6.1.

Corollary 6.1. *Suppose $f_H \in \mathcal{L}$, $n \geq 0$, and $\rho(A_H) < 1$. Then*

$$\mu(\mathbb{T}^n(f_H)) = \mathbf{1}(I - A_H)^{-1}(I - A_H^{n+1})\mathbf{e}_H; \text{ and}$$

$$\lim_{n \to \infty} \mu(\mathbb{T}^n(f_H)) = \mathbf{1}(I - A_H)^{-1}\mathbf{e}_H,$$

where I is the identity matrix.

Proof. If $\lambda \in \sigma(A_H)$, then $A_H \mathbf{v} = \lambda \mathbf{v}$ for some $\mathbf{v} \in \mathbb{C}^m$. Hence $(I - A_H)\mathbf{v} = (1 - \lambda)\mathbf{v}$, implying $\sigma(I - A_H) = \{1 - \lambda : \lambda \in \sigma(A_H)\}$. If $\rho(A_H) < 1$, then it follows that $0 \notin \sigma(I - A_H)$, so that $I - A_H$ is invertible.

The equalities in the corollary follow from those in Theorem 6.1 together with the identity

$$\sum_{i=0}^{n} A_H^i = (I - A_H)^{-1}(I - A_H^{n+1}),$$

since for every matrix $M \in \mathbb{C}^{m \times m}$, we have $\lim_{n \to \infty} M^n = 0$ if and only if $\rho(M) < 1$. □

Corollary 6.2. *Suppose $f_H \in \mathcal{L}$ with $H = \cup_{i=1}^{m} \xi_i$. If $\rho(A_H) = 0$, then $\mu(\mathbb{T}^n(f_H)) = 1$ for some $n < \infty$.*

Proof. If $\rho(A_H) = 0$ then the graph Γ_H has no cycles. Therefore, if $x \in \xi_i \not\subset H$ then $f^k(x) \notin x_i$ for all $k > 0$. Since $\xi = \cup_{i=1}^{m} \xi_i$ is a finite Markov partition then $f^m(x) \in H$ for all $x \in I$. □

The matrix $M \in \mathbb{C}^{m \times m}$ is called *defective* if it does not have an eigenbasis, i.e., if there are not enough linearly independent eigenvectors of M to form a basis of $\mathbb{C}^m$. A matrix with an eigenbasis is called *nondefective*.

Corollary 6.3. *Let $f_H \in \mathcal{L}$ and suppose the matrix A_H is nondefective with eigenpairs $\{(\lambda_1, \mathbf{v}_1), \ldots, (\lambda_m, \mathbf{v}_m)\}$ with no eigenvalue equal to 1. Then $\mathbf{e}_H = \sum_{i=1}^{m} c_i \mathbf{v}_i$ for some $c_1, \ldots, c_m \in \mathbb{C}$ and*

$$\mu(\mathbb{E}^n(f_H)) = \sum_{i=1}^{m} c_i s_i \lambda_i^n \tag{6.5}$$

$$\mu(\mathbb{T}^n(f_H)) = \sum_{i=1}^{m} c_i s_i \left(\frac{1 - \lambda_i^{n+1}}{1 - \lambda_i} \right), \tag{6.6}$$

where $s_i = \mathbf{1}\mathbf{v}_i$.

Proof. Equations (6.5) and (6.6) follow from Theorem 6.1 together with the identity $\sum_{k=0}^{m} \lambda_i^k = (1 - \lambda_i)(1 - \lambda_i^{n+1})$, since it is assumed that $\lambda_i \neq 1$ for each $1 \leq i \leq m$. □

Example 6.2. Let the function $f : I \to I$ be the tent map considered in Example 6.3 and let $H = (0, 1/4]$. Since $f_H \in \mathcal{L}$, one can calculate that f_H has the weighted transition matrix

$$A_H = \begin{bmatrix} 0 & 0 & 0 & 0 \\ 0 & 0 & 1/2 & 1/2 \\ 0 & 0 & 1/2 & 1/2 \\ 1/2 & 1/2 & 0 & 0 \end{bmatrix}.$$

6.2 Piecewise Linear Functions

The matrix A_H is nondefective, since its eigenvalues $\sigma(A_H) = \{\frac{1+\sqrt{5}}{4}, \frac{1-\sqrt{5}}{4}, 0, 0\}$ correspond to the linearly independent eigenvectors

$$\mathbf{v}_1 = \begin{bmatrix} 0 \\ \frac{1+\sqrt{5}}{4} \\ \frac{1+\sqrt{5}}{4} \\ 1 \end{bmatrix}, \mathbf{v}_2 = \begin{bmatrix} 0 \\ \frac{1-\sqrt{5}}{4} \\ \frac{1-\sqrt{5}}{4} \\ 1 \end{bmatrix}, \mathbf{v}_3 = \begin{bmatrix} 0 \\ 0 \\ -1 \\ 1 \end{bmatrix}, \mathbf{v}_4 = \begin{bmatrix} -1 \\ 1 \\ 0 \\ 0 \end{bmatrix},$$

respectively. Since the vector $\mathbf{e}_H = [1/4, 0, 0, 0]^T$ can be written as

$$\mathbf{e}_H = \frac{5-\sqrt{5}}{20+20\sqrt{5}} \mathbf{v}_1 - \frac{3+\sqrt{5}}{8\sqrt{5}} \mathbf{v}_2 + \frac{1}{4} \mathbf{v}_3 - \frac{1}{4} \mathbf{v}_4,$$

equations (6.5) and (6.6) in Corollary 6.3 imply

$$\mathbb{E}^n(f_H) = \frac{1}{40}(5+\sqrt{5})\lambda_1^n + \frac{1}{40}(5-\sqrt{5})\lambda_2^n; \text{ and}$$

$$\mathbb{T}^n(f_H) = \frac{1}{40}(5+\sqrt{5})\frac{1-\lambda_1^{n+1}}{1-\lambda_1} + \frac{1}{40}(5-\sqrt{5})\frac{1-\lambda_2^{n+1}}{1-\lambda_2}$$

$$= 1 - \left(\frac{1}{2} + \frac{1}{\sqrt{5}}\right)\lambda_1^{n+1} - \left(\frac{1}{2} - \frac{1}{\sqrt{5}}\right)\lambda_2^{n+1}.$$

Since $\sigma(A_H) < 1$, it follows that by computing the matrix $(I - A_H)^{-1}$, we obtain

$$\mathbf{1}(I - A_H)^{-1}\mathbf{e}_H = [1\ 1\ 1\ 1] \begin{bmatrix} 1 & 0 & 0 & 0 \\ 1 & 2 & 2 & 2 \\ 1 & 1 & 3 & 2 \\ 1 & 1 & 1 & 2 \end{bmatrix} \begin{bmatrix} 1/4 \\ 0 \\ 0 \\ 0 \end{bmatrix} = 1.$$

Corollary 6.1 then implies that $\lim_{n\to\infty} \mathbb{T}^n(f_H) = 1$. Hence, the probability $\mathbb{P}(f_H)$ of surviving indefinitely in this system, for a typical $x \in I$, is in fact zero. This can be seen in Fig. 6.2, where both $\mu(\mathbb{E}^n(f_H))$ and $\mu(\mathbb{T}^n(f_H))$ are plotted.

For a fixed map $f : I \to I$, it is possible to consider the two open systems (f_J, I) and (f_K, I), where J and K are two different holes. One of the questions Theorem 6.1 allows us to answer is whether $\mu(\mathbb{E}^n(f_J)) \geq \mu(\mathbb{E}^n(f_K))$ or $\mu(\mathbb{T}^n(f_J)) \geq \mu(\mathbb{T}^n(f_K))$ for every $n \geq 0$. Stated less formally, we can determine which hole is leaking the most at any moment in time. This particular topic is considered in more detail in [2].

If the holes J and K are disjoint, another question we can address is whether more phase space is escaping from J than from K in the open system $f_{J \cup K}$. This is quite different from asking whether more phase space escapes from f_J at time n than from f_K. The reason is that if we simultaneously have the holes J and K,

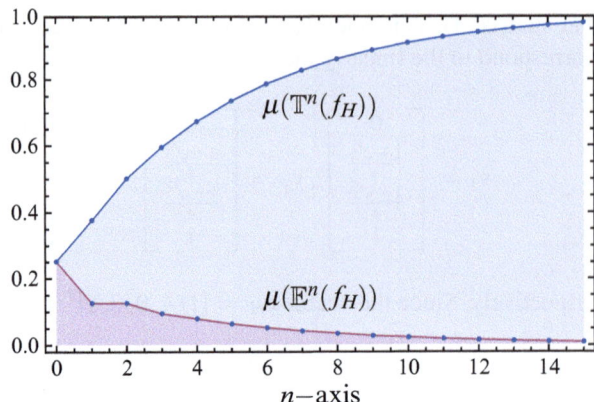

Fig. 6.2 Plots of $\mu(\mathbb{E}^n(f_H))$ and $\mu(\mathbb{T}^n(f_H))$ for the open system (f_H, I) in Example 6.2

then it may happen that the phase space that would have escaped through J at time n in the absence of K could already have escaped through K at some time $k < n$. In this way, K may "block" escape through J, and J may block escape through K. Answering the question through which hole more phase space escapes involves taking into account this blocking process.

Suppose $f : I \to I$ has the Markov partition $\xi = \cup_{i=1}^{m} \xi_i$ with holes $H_1, \ldots, H_\ell$. Assuming $H = \cup_{i=1}^{\ell} H_i$, where the holes $H_1 \ldots, H_\ell$ are disjoint, we define

$$\mathbb{E}_i^n(f_H) = \{x \in I : f_H^n(x) \in H_i, \; f_H^k(x) \notin H_i, \; 0 \leq k < n\}; \text{ and}$$

$$\mathbb{T}_i^n(f_H) = \{x \in I : f_H^k(x) \in H_i, \text{ for some } k, \; 0 \leq k \leq n\};$$

for each $1 \leq i \leq \ell$ and $n \geq 0$. The set $\mathbb{E}_i^n(f_H)$ consists of those points that escape through H_i at time n, while $\mathbb{T}_i^n(f_H)$ is the set of points that escape through H_i before time $n+1$ in the open system $f_H : I \to I$. Hence,

$$\bigcup_{i=1}^{\ell} \mathbb{E}_i^n(f_H) = \mathbb{E}^n(f_H) \text{ and } \bigcup_{i=1}^{\ell} \mathbb{T}_i^n(f_H) = \mathbb{T}^n(f_H).$$

If $H_i = \cup_{j \in \mathscr{I}} \xi_j$, then we let the vector $\mathbf{e}_{H_i}$ be given by

$$(\mathbf{e}_{H_i})_j = \begin{cases} \mu(\xi_j) & \text{if } j \in \mathscr{I}, \\ 0 & \text{otherwise.} \end{cases}$$

The following result is a direct consequence of the proof of Theorem 6.1.

6.2 Piecewise Linear Functions

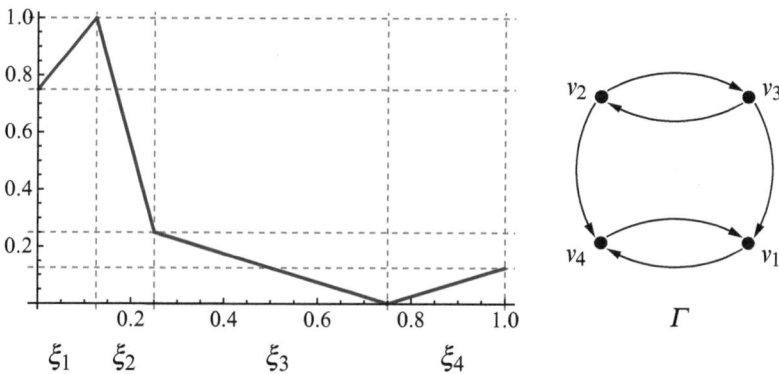

Fig. 6.3 The transition graph Γ (right) of the closed system $f : I \to I$ (left) in Example 6.3

Corollary 6.4. *Let $f_H \in \mathscr{L}$, where $H = \cup_{i=1}^{\ell} H_i$ and $H_1, \ldots, H_\ell$ are disjoint. Then*

$$\mu(\mathbb{E}_i^n(f_H)) = \mathbf{1} A_H^n \mathbf{e}_{H_i}; \text{ and}$$

$$\mu(\mathbb{T}_i^n(f_H)) = \mathbf{1} \left(\sum_{j=0}^{n} A_H^j \right) \mathbf{e}_{H_i}$$

for each $1 \leq i \leq \ell$ and $n \geq 0$.

Example 6.3. Let the function $f : I \to I$ be the map

$$f(x) = \begin{cases} 2x + 3/4 & 0 \leq x \leq 1/8, \\ -6x + 7/4 & 1/8 \leq x \leq 1/4, \\ -(1/2)x + 3/8 & 1/4 \leq x \leq 3/4, \\ (1/2)x - 3/8 & 3/4 < x \leq 1, \end{cases}$$

with Markov partition $\xi = \{(0, 1/8], (1/8, 1/4], (1/4, 3/4], (3/4, 1]\}$. This map is shown in Fig. 6.3 (left) along with its graph of transitions Γ (right). Here, we consider the holes $H_1 = (0, 1/4]$, $H_2 = (3/4, 1]$, and $H = H_1 \cup H_2$. We note that $f_{H_1}, f_{H_2}, f_H \in \mathscr{L}$, so that each open system has a graph of transitions. These are shown in Fig. 6.4 (bottom), at the left, right, and center, respectively.

Beginning with the holes H_1 and H_2, the open systems (f_{H_1}, I) and (f_{H_2}, I) have the weighted transition matrices given by

$$A_{H_1} = \begin{bmatrix} 0 & 0 & 0 & 0 \\ 0 & 0 & 1/6 & 1/6 \\ 2 & 2 & 0 & 0 \\ 2 & 0 & 0 & 0 \end{bmatrix} \text{ and } A_{H_2} = \begin{bmatrix} 0 & 0 & 0 & 1/2 \\ 0 & 0 & 1/6 & 1/6 \\ 2 & 2 & 0 & 0 \\ 0 & 0 & 0 & 0 \end{bmatrix}$$

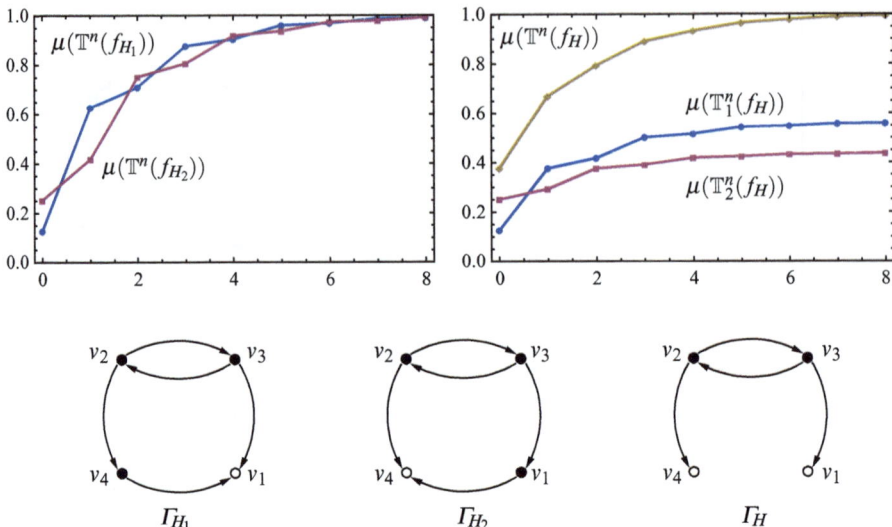

Fig. 6.4 The measure of the phase space that has escaped in f_{H_1} (*blue*) and f_{H_2} (*red*) is shown (*top left*). The measure of the phase space that escapes through the holes H_1, H_2, and H in f_H is shown (*top right*). The transition graphs Γ_{H_1}, Γ_{H_2}, and Γ_H are shown *below*

respectively. Using Theorem 6.1, both $\mu(\mathbb{E}^n(f_{H_1}))$ and $\mu(\mathbb{T}^n(f_{H_2}))$ can be found for all $n \leq 0$. These are shown in Fig. 6.4 (top left) in blue and red, respectively. We note that in both cases,

$$\lim_{n\to\infty} \mathbb{P}^n(f_{H_1}) = 0 \text{ and } \lim_{n\to\infty} \mathbb{P}^n(f_{H_2}) = 0,$$

so that almost every point eventually escapes from both these systems.

The system f_H, with both holes H_1 and H_2, has the weighted transition matrix

$$A_H = \begin{bmatrix} 0 & 0 & 0 & 0 \\ 0 & 0 & 1/6 & 1/6 \\ 2 & 2 & 0 & 0 \\ 0 & 0 & 0 & 0 \end{bmatrix}.$$

Using Theorem 6.1 and Corollary 6.4, it is possible to compute each of $\mu(\mathbb{T}_1^n(f_H))$, $\mu(\mathbb{T}_2^n(f_H))$, and $\mu(\mathbb{T}^n(f_H))$. These are shown in Fig. 6.4 (top right), where we note that $\mu(\mathbb{T}^n(f_H)) = \mu(\mathbb{T}_1^n(f_H)) + \mu(\mathbb{T}_2^n(f_H))$ for $n \geq 0$. However, the measure of points that escape through H_1 by time n is larger than the measure of those points that escape through H_2 for all $n > 0$.

This is interesting for the following reason. In the systems f_{H_1} and f_{H_2}, almost every point escapes through these holes separately at nearly the same rate. However, in the open system f_H, different amounts of phase space escape the system through H_1 and H_2 respectively. In particular,

$$\lim_{n\to\infty} \mu\big(\mathbb{T}_1^n(f_H)\big) = 0.5625 \quad \text{and} \quad \lim_{n\to\infty} \mu\big(\mathbb{T}_2^n(f_H)\big) = 0.4375.$$

One way to state this is to say that the hole H_1 does a better job of blocking phase space from entering H_2 than H_2 does from keeping points from escaping through H_1. This may seem surprising when one considers the fact that H_1 has a smaller Lebesgue measure than H_2.

We note that this idea of simultaneously open holes where $H = \cup_{i=1}^\ell H_i$ is quite natural, since every hole H in our setup is composed of a number of Markov partition elements $H = \cup_{i \in \mathscr{I}} \xi_i$. Since each partition element ξ_i or collection of such elements can be thought of as a hole, Corollary 6.4 allows us to determine which *part* of the hole H is leaking the most.

6.3 Nonlinear Estimates

We now consider the open systems $f_H : I \to I$, where f is allowed to be a nonlinear but differentiable function when restricted to the elements of ξ. The formulas we derive in this section allow us to give upper and lower bounds on $\mu(\mathbb{E}^n(f_H))$ and $\mu(\mathbb{T}^n(f_H))$ for every finite time $n \geq 0$. In particular, the finite time estimates of $\mu(\mathbb{T}^n(f_H))$ can be used to bound the asymptotic survival probability $\mathbb{P}(f_H)$.

Suppose $H = \bigcup_{i \in \mathscr{I}} \xi_i$ for some $\mathscr{I} \subset M$. Let $\mathscr{N}$ be the open systems $f_H : I \to I$, which have the property that

$$\inf_{x \in \xi_i} |f_H'(x)| > 0 \text{ for } i \notin \mathscr{I}; \text{ and}$$

$$\sup_{x \in \xi_i} |f_H'(x)| < \infty \text{ for } i \notin \mathscr{I}.$$

To each $f_H \in \mathscr{N}$ there are two associated matrices similar to the weighted transition matrix A_H defined for each open system in $\mathscr{L}$.

Definition 6.4. Suppose $f_H \in \mathscr{N}$, where $H = \bigcup_{i \in \mathscr{I}} \xi_i$ for some $\mathscr{I} \subset M$. The matrix $\underline{A}_H \in \mathbb{R}^{m \times m}$ is defined by

$$(\underline{A}_H)_{ij} = \begin{cases} \inf_{x \in \xi_{ij}} |f'(x)|^{-1} & \text{for } \xi_{ij} \neq \emptyset\ i \notin \mathscr{I}, \\ 0 & \text{otherwise,} \end{cases} \quad 1 \leq i, j \leq m.$$

Similarly, the matrix $\overline{A}_H \in \mathbb{R}^{m \times m}$ is defined by

$$(\overline{A}_H)_{ij} = \begin{cases} \sup_{x \in \xi_{ij}} |f'(x)|^{-1} & \text{for } \xi_{ij} \neq \emptyset, \ i \notin \mathscr{I}, \\ 0 & \text{otherwise,} \end{cases} \quad 1 \leq i, j \leq m.$$

For $f_H \in \mathscr{N}$ and $n \geq 0$, let

$$\underline{\mathbb{E}}^n(f_H) = \mathbf{1}\underline{A}_H^n \mathbf{e}_H \text{ and } \overline{\mathbb{E}}^n(f_H) = \mathbf{1}\overline{A}_H^n \mathbf{e}_H;$$

$$\underline{\mathbb{T}}^n(f_H) = \mathbf{1}\left(\sum_{i=0}^{n} \underline{A}_H^i\right)\mathbf{e}_H \text{ and } \overline{\mathbb{T}}^n(f_H) = \mathbf{1}\left(\sum_{i=0}^{n} \overline{A}_H^i\right)\mathbf{e}_H.$$

Since the function $f_H : I \to I$ may be nonlinear on the elements of ξ, it is not possible to give an exact formula for either $\mathbb{E}^n(f_H)$ or $\mathbb{T}^n(f_H)$, as in the previous section. Instead, here we give bounds on both of these quantities.

Theorem 6.2. *If $f_H \in \mathscr{N}$ and $n \geq 0$, then*

$$\underline{\mathbb{E}}^n(f_H) \leq \mu(\mathbb{E}^n(f_H)) \leq \overline{\mathbb{E}}^n(f_H); \text{ and} \tag{6.7}$$

$$\underline{\mathbb{T}}^n(f_H) \leq \mu(\mathbb{T}^n(f_H)) \leq \overline{\mathbb{T}}^n(f_H). \tag{6.8}$$

Proof. Similar to the proof of Theorem 6.1, let

$$\xi_{i_0 i_1 \ldots i_n} = \{x \in I : f^k(x) \in \xi_{i_k} \text{ for } 0 \leq k \leq n\},$$

where $f_H : I \to I$ has the Markov partition $\xi = \cup_{i=1}^{m} \xi_i$ and $H = \cup_{i \in \mathscr{I}} \xi_i$ for some $\mathscr{I} \subseteq M$. Additionally, let $\underline{z}_i = \inf_{x \in \xi_i} |f'(x)|^{-1}$. Assuming $\xi_{i_0 i_1} \neq \emptyset$, then $\mu(\xi_{i_0 i_1}) \geq \underline{z}_{i_0} \mu(\xi_{i_1})$. Continuing inductively, it follows that if $\xi_{i_0 i_1 \ldots i_n} \neq \emptyset$, then

$$\mu(\xi_{i_0 i_1 \ldots i_n}) \geq \left(\prod_{k=0}^{n-1} \underline{z}_{i_k}\right) \mu(\xi_{i_n}). \tag{6.9}$$

By definition, the measure of the subset of I that escapes the system at time n is

$$\mu(\mathbb{E}^n(f_H)) = \sum_{\gamma} \mu(\xi_\gamma) = \sum_{\ell=1}^{m}\left(\sum_{\gamma_\ell} \mu(\xi_{\gamma_\ell})\right), \tag{6.10}$$

where the last sum is taken over all sequences $\gamma_\ell = i_0 i_1 \ldots i_n$ in which $i_0 = \ell$, $\xi_{i_k} \not\subset H$ for $0 \leq k < n$, $\xi_{i_n} \subset H$, and $\xi_{\gamma_\ell} \neq \emptyset$.

6.3 Nonlinear Estimates

Following the proof of Theorem 6.1, we have that

$$(\underline{A}_H^n \mathbf{e}_H)_\ell = \sum_{\gamma_\ell} \left(\prod_{k=0}^{n-1} (\underline{A}_H)_{i_k i_{k+1}} \right) (\mathbf{e}_H)_{i_n}, \tag{6.11}$$

where the sum is taken over all $\gamma_\ell = i_0 i_1 \ldots i_n$. Here, similar to (6.10), the index i_0 is equal to ℓ, $\xi_{i_k} \not\subset H$ for $0 \le k < n$, $\xi_{i_n} \subset H$, and $\xi_{\gamma_\ell} \ne \emptyset$.

Since $(\underline{A}_H)_{i_k i_{k+1}} = \underline{z}_{i_k}$ for each $0 \le k < n$, under the assumption that $\xi_{\gamma_\ell} \ne \emptyset$, it follows from (6.9) and (6.11) that

$$(\underline{A}_H^n \mathbf{e}_H)_\ell = \sum_{\gamma_\ell} \left(\prod_{k=0}^{n-1} \underline{z}_{i_k} \right) \mu(\xi_{i_n}) \le \sum_{\gamma_\ell} \mu(\xi_{\gamma_\ell}).$$

By use of (6.10), we have

$$\mathbf{1}\underline{A}_H^n \mathbf{e}_H = \sum_{\ell=1}^m \left(\sum_{\gamma_\ell} \left(\prod_{k=0}^{n-1} \underline{z}_{i_k} \right) \mu(\xi_{i_n}) \right) \le \sum_{\ell=1}^m \left(\sum_{\ell=1}^m \mu(\xi_{\gamma_\ell}) \right) = \mu(\underline{\mathbb{E}}^n(f_H)).$$

Similarly, one can show that $\mathbf{1}\overline{A}_H^n \mathbf{e}_H \ge \mu(\overline{\mathbb{E}}^n(f_H))$. It then follows from the inequalities $\underline{\mathbb{E}}^n(f_H) \le \mu(\mathbb{T}^n(f_H)) \le \overline{\mathbb{E}}^n(f_H)$ that $\underline{\mathbb{T}}^n(f_H) \le \mu(\mathbb{T}^n(f_H)) \le \overline{\mathbb{T}}^n(f_H)$. This completes the proof. $\square$

Theorem 6.2 allows us to bound the measure of those points that escape through H at time n and before time $n+1$ in the open system $f_H : I \to I$. If the matrices $\underline{A}_H$ and $\overline{A}_H$ are nondefective, then we have the following result similar to Corollary 6.3.

Corollary 6.5. *Let $f_H \in \mathcal{N}$ and suppose both $\underline{A}_H$ and $\overline{A}_H$ are nondefective with eigenpairs $\{(\underline{\lambda}_1, \underline{\mathbf{v}}_1), \ldots, (\underline{\lambda}_k, \underline{\mathbf{v}}_k)\}$ and $\{(_1, \overline{\mathbf{v}}_1), \ldots, (\overline{\lambda}_k, \overline{\mathbf{v}}_k)\}$ respectively, where no eigenvalue is equal to 1. Then for each $n \ge 0$,*

$$\sum_{i=1}^k \underline{c}_i \underline{s}_i \underline{\lambda}_i^n \le \mu(\underline{\mathbb{E}}^n(f_H)) \le \sum_{i=1}^k \overline{c}_i \overline{s}_i \overline{\lambda}_i^n; \text{ and} \tag{6.12}$$

$$\sum_{i=1}^k \underline{c}_i \underline{s}_i \left(\frac{1 - \underline{\lambda}_i^{n+1}}{1 - \underline{\lambda}_i} \right) \le \mu(\mathbb{T}^n(f_H)) \le \sum_{i=1}^k \overline{c}_i \overline{s}_i \left(\frac{1 - \overline{\lambda}_i^{n+1}}{1 - \overline{\lambda}_i} \right), \tag{6.13}$$

where $\underline{s}_i = \mathbf{1}\underline{\mathbf{v}}_i$, $\overline{s}_i = \mathbf{1}\overline{\mathbf{v}}_i$, $\mathbf{e}_H = \sum_{i=1}^n \underline{c}_i \underline{\mathbf{v}}_i$, and $\mathbf{e}_H = \sum_{i=1}^n \overline{c}_i \overline{\mathbf{v}}_i$.

The proof of Corollary 6.5 is analogous to the proof of Corollary 6.3. Also, the upper and lower bounds given in (6.12) are the quantities $\underline{\mathbb{E}}^n(f_H)$ and $\overline{\mathbb{E}}^n(f_H)$,

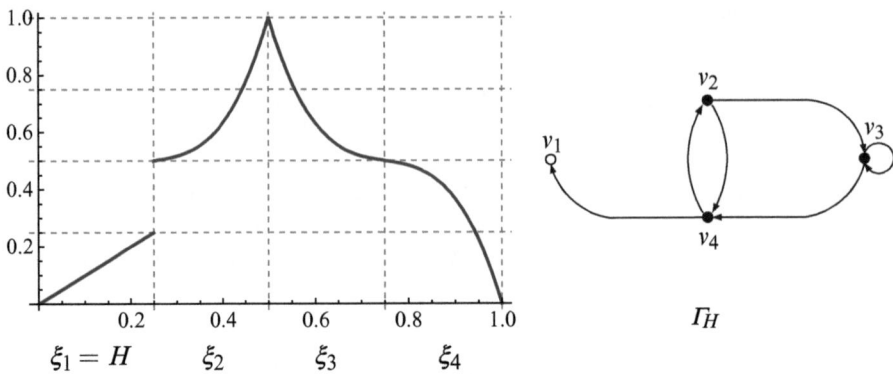

Fig. 6.5 The transition graph Γ_H (right) of the open system $g_H : I \to I$ (left) in Example 6.4

respectively. Moreover, the upper and lower bounds given in (6.13) are $\underline{\mathbb{T}}^n(f_H)$ and $\overline{\mathbb{T}}^n(f_H)$, respectively.

Example 6.4. Consider the function $g : I \to I$ given by

$$g(x) = \begin{cases} \frac{11}{2}x - 21x^2 + 28x^3, & 0 \le x \le 1/2, \\ \frac{11}{2}(1-x) - 21(1-x)^2 + 28(1-x)^3, & 1/2 < x \le 1, \end{cases}$$

with Markov partition $\xi = \{(0, 1/4], (1/4, 1/2], (1/2, 3/4], (3/4, 1]\}$ and $H = (0, 1/4]$. The function $g : I \to I$ can be considered a nonlinear version of the tent map $f : I \to I$ in Example 6.3.

Here, as in Example 6.3, we let $H = (0, 1/4]$. For the open system (g_H, I), one can compute that $\xi_{23} = (1/4, 0.44]$, $\xi_{24} = (0.44, 1/2]$, $\xi_{34} = (1/2, 0.55]$, $\xi_{33} = (0.55, 0.3/4]$, $\xi_{42} = (3/4, 0.94]$, and $\xi_{41} = (0.94, 1]$. From this, we find that

$$\underline{A}_H = \begin{bmatrix} 0 & 0 & 0 & 0 \\ 0 & 0 & 0.29 & 2/11 \\ 0 & 0 & 0.29 & 2/11 \\ 2/11 & 0.29 & 0 & 0 \end{bmatrix} \text{ and } \overline{A}_H = \begin{bmatrix} 0 & 0 & 0 & 0 \\ 0 & 0 & 4 & 0.29 \\ 0 & 0 & 4 & 0.29 \\ 0.29 & 4 & 0 & 0 \end{bmatrix}.$$

Since $\underline{A}_H$ and $\overline{A}_H$ are nondefective, one can compute using Corollary 6.5 that

$$0.15\underline{\lambda}_1^n + 0.15\underline{\lambda}_2^n - 0.6 \le \mathbb{E}^n(g_H) \le -0.02\overline{\lambda}_1^n + 0.02\overline{\lambda}_2^n + 0.25, \quad (6.14)$$

where $\underline{\lambda}_1 = 0.41$, $\underline{\lambda}_2 = -0.12$, $\overline{\lambda}_1 = 4.27$, and $\overline{\lambda}_2 = -0.27$. By plotting the inequalities in (6.14), we have the graphs shown in Fig. 6.5. Here the shaded area indicates the region in which $\mu(\mathbb{E}^n(g_H))$ must lie.

6.4 Improved Escape Estimates

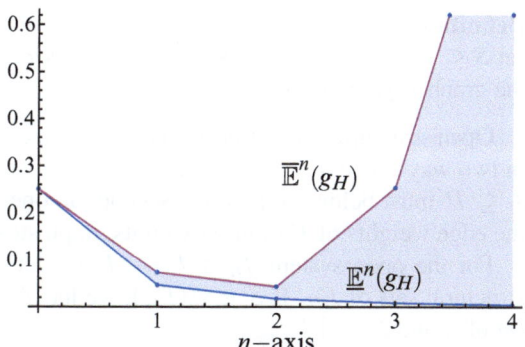

Fig. 6.6 The *upper bounds* $\overline{\mathbb{E}}^n(g_H)$ and *lower bounds* $\underline{\mathbb{E}}^n(g_H)$ of $\mu(\mathbb{E}^n(g_H))$ are shown for the open system $g_H : I \to I$ considered in Example 6.4

Suppose $f_H \in \mathcal{N}$, in which H is made up of a number of disjoint holes $H_1, \ldots, H_\ell$. We can again ask through which one of these holes escape is the fastest. To answer this, we let

$$\underline{\mathbb{E}}^n_i(f_H) = \mathbf{1}\underline{A}^n_H \mathbf{e}_{H_i} \text{ and } \overline{\mathbb{E}}^n_i(f_H) = \mathbf{1}\overline{A}^n_H \mathbf{e}_{H_i};$$

$$\underline{\mathbb{T}}^n_i(f_H) = \mathbf{1}\left(\sum_{i=0}^n \underline{A}^i_H\right)\mathbf{e}_{H_i} \text{ and } \overline{\mathbb{T}}^n_i(f_H) = \mathbf{1}\left(\sum_{i=0}^n \overline{A}^i_H\right)\mathbf{e}_{H_i}.$$

As a direct consequence of the proof of Theorem 6.2, we have the following result.

Corollary 6.6. *Let $f_H \in \mathcal{N}$, where $H = \cup_{i=1}^\ell H_i$ and $H_1, \ldots, H_\ell$ are disjoint. Then*

$$\underline{\mathbb{E}}^n_i(f_H) \leq \mu(\mathbb{E}^n_i(f_H)) \leq \overline{\mathbb{E}}^n_i(f_H); \text{ and}$$

$$\underline{\mathbb{T}}^n_i(f_H) \leq \mu(\mathbb{T}^n_i(f_H)) \leq \overline{\mathbb{T}}^n_i(f_H),$$

for each $1 \leq i \leq \ell$ and $n \geq 0$.

Similarly to Corollary 6.4, Corollary 6.6 allows us to bound the Lebesgue measure of the set that escapes through a specific part H.

We now turn our attention to finding ways of improving our estimates of $\mathbb{E}^n(f_H)$ and $\mathbb{T}^n(f_H)$, found in Theorem 6.2.

6.4 Improved Escape Estimates

In this section, we define a delayed first return map of an open system $f_H \in \mathcal{N}$, which we will use to improve the escape estimates given in Theorem 6.2. A key step in this procedure is to choose a particular vertex set of Γ_H over which this map will be defined.

Definition 6.5. Let $H = \bigcup_{i \in \mathscr{I}} \xi_i$ for some $\mathscr{I} \subset M$ and let $\Gamma_H = (V, E_H)$. The set $S \subseteq V = \{v_1, \ldots, v_m\}$ is an *open structural set* of Γ_H if $v_i \in S$ for $i \in \mathscr{I}$, and the graph $\Gamma_H|_{\bar{S}}$ has no cycles.

Open structural sets differ from the complete structural sets introduced in Chap. 2 in two ways. The first is that each vertex v_i that corresponds to a partition element $\xi_i \subseteq H$ must belong to this set. Second, an open structural set does not depend on the edge weights of Γ_H but only on its graph structure.

For the open system $f_H : I \to I$, we let $st(\Gamma_H)$ denote the set of all open structural sets of Γ_H. If $S \in st(\Gamma_H)$, we let $\mathscr{I}_S = \{i \in M : v_i \in S\}$ be the *index set* of S and $\xi_S = \bigcup_{i \in \mathscr{I}_S} \xi_i$.

Definition 6.6. Let $S \in st(\Gamma_H)$. For $x \in I$, we let $\gamma(x) = i_0 i_1 \ldots i_t$, where $i_j = k$ if $f_H^j(x) \in \xi_k$ and t is the smallest $k > 0$ such that $f_H^k(x) \in \xi_S$. The set

$$\Omega_S = \{\gamma : \gamma = \gamma(x) \text{ for some } x \in I \setminus H\}$$

consists of the *admissible sequences* of f_H with respect to S.

For $x \in I$, we say that $\gamma(x) = i_0 i_1 \ldots i_t$ has length $|\gamma(x)| = t$. The reason that we have $|\gamma(x)| < \infty$ is that the graph $\Gamma_H|_{\bar{S}}$ has no cycles. Hence, after a finite number of steps, $f_H^t(x)$ must enter ξ_S.

Definition 6.7. Let $S \in st(\Gamma_H)$. For $x_0 \in I$ and $k \geq 0$, we inductively define the function $x_{k+1} = Rf_S(x_k, \ldots, x_0)$, where

$$x_{k+1} = \begin{cases} f_H^{|\gamma(x_k)|}(x_k) & \text{if } x_{k-i} = x_k \text{ for each } 0 \leq i \leq |\gamma(x_k)| - 1, \\ x_k & \text{otherwise.} \end{cases}$$

The function $Rf_S : I^{k+1} \to I$ is called the *delayed first return map* of f_H with respect to S. The sequence $x_0, x_1, x_2, \ldots$ is the *orbit* of x_0 under Rf_S.

If $T = \max_{x \in I} |\gamma(x)|$, then strictly speaking, the map $x_{k+1} = Rf_S(x_k, \ldots, x_\tau)$ for some $\tau < T$. As can be seen, the map Rf_S acts almost like a first return map of f_H to the set ξ_S. The difference is that a return to ξ_S does not happen instantaneously (as would happen in the case of a first return map) but is delayed, so that the trajectory of a point under f_H coincides with the trajectory under Rf_S after a return to ξ_S.

We say that the map $Rf_S : I^{k+1} \to I$ generates the open dynamical system (Rf_S, I). For $n \geq 0$, we let $Rf_S^n(x_0) = x_n$ and define the sets

$$\mathbb{E}^n(Rf_S) = \{x \in I : Rf_S^n(x) \in H, \ Rf_S^k(x) \notin H, \ 0 \leq k < n\}; \text{ and}$$

$$\mathbb{T}^n(Rf_S) = \{x \in I : Rf_S^k(x) \in H, \text{ for some } k, \ 0 \leq k \leq n\}.$$

Lemma 6.1. *If $S \in st_\xi(\Gamma_{f_H})$ and $n \geq 0$, then $\mathbb{E}^n(f_H) = \mathbb{E}^n(Rf_S)$.*

6.4 Improved Escape Estimates

Proof. For $x_0 \in I$, let $\tilde{\gamma}(x_0) = i_0 i_1 \ldots$, where $i_j = k$ if $f_H^j(x_0) \in \xi_k$. Choosing $S \in st(\Gamma_H)$, let $\tilde{\gamma}_S(x_0) = \ell_0 \ell_1 \ldots$, where $\ell_j = k$ if $Rf_S^j(x_0) \in \xi_k$. Let $t > 0$ be the smallest number such that $i_t \in \mathscr{I}_S$. Then $\gamma(x_0) = i_0 i_1 \ldots i_t$, and Definition 6.7 implies that $Rf_S^t(x_0) = f_H^t(x_0)$. Therefore, $i_t = \ell_t$ where $t \in \mathscr{I}_S$.

Continuing in this manner, it follows that $i_j = \ell_j$ for each $j \in \mathscr{I}_S$. Since $H \subseteq \xi_S$, the point x_0, if it escapes, will escape for both f_H and Rf_S at exactly the same time. This completes the proof. □

As a consequence of Lemma 6.1, one can use (Rf_S, I) to study the escape in the open system (f_H, I). However, because of the time delays involved, the weighted transition matrix of (Rf_S, I) cannot be defined in the same way that we have defined $\underline{A}_H$ and $\overline{A}_H$.

To define a transition matrix of (Rf_S, I), we need the following. For $S \in st(\Gamma_H)$, let

$$M_S = M \cup \{\gamma; i : \gamma \in \Omega_S, 0 < i < |\gamma|\}.$$

If $\gamma = i_0 \ldots, i_t$, we identify the index $\gamma; 0$ with i_0 and the index $\gamma; t$ with i_t. We also let $\xi_\gamma = \{x \in I : \gamma(x) = \gamma\}$ for each admissible sequence $\gamma \in \Omega_S$, which simply extends our notation given by (6.1) in Sect. 6.2.

Definition 6.8. For $S \in st(\Gamma_H)$, let $\underline{A}_S$ be the matrix with rows and columns indexed by elements of M_S, where

$$(\underline{A}_S)_{ij} = \begin{cases} \inf_{x \in \xi_\gamma} |(f^{|\gamma|}(x))'|^{-1} & \text{if } i = \gamma; |\gamma| - 1, \ j = \gamma; |\gamma|, \text{ for some, } \gamma \in \Omega_S, \\ 1 & \text{if } i = \gamma; k - 1, \ j = \gamma; k, \ k \neq |\gamma|, \text{ for some, } \gamma \in \Omega_S, \\ 0 & \text{otherwise.} \end{cases}$$

(6.15)

We call $\underline{A}_S$ the *lower transition matrix* of Rf_S. The matrix $\overline{A}_S$ defined by replacing the infimum in (6.15) by the supremum is the *upper transition matrix* of Rf_S.

Let $\mathbf{1}_S$ be the $1 \times |M_S|$ vector given by

$$(\mathbf{1}_S)_i = \begin{cases} 1 & \text{if } i \in M, \\ 0 & \text{otherwise.} \end{cases}$$

Let $\mathbf{e}_S$ be the $|M_S| \times 1$ vector given by

$$(\mathbf{e}_S)_i = \begin{cases} \mu(\xi_i) & \text{if } i \in \mathscr{I}, \\ 0 & \text{otherwise.} \end{cases}$$

Lastly, for $n \geq 0$, let

$$\underline{\mathbb{E}}^n(Rf_S) = \mathbf{1}_S \underline{A}_S^n \mathbf{e}_S \text{ and } \overline{\mathbb{E}}^n(Rf_S) = \mathbf{1}_S \overline{A}_S^n \mathbf{e}_S;$$

$$\underline{\mathbb{T}}^n(Rf_S) = \mathbf{1}_S \left(\sum_{i=0}^n \underline{A}_S^i \right) \mathbf{e}_S \text{ and } \overline{\mathbb{T}}^n(Rf_S) = \mathbf{1}_S \left(\sum_{i=0}^n \overline{A}_S^i \right) \mathbf{e}_S.$$

Using these quantities, we give the following improved escape estimates.

Theorem 6.3. *Let $f_H \in \mathcal{N}$ and suppose $S \in st(\Gamma_H)$. If $n \geq 0$, then*

$$\underline{\mathbb{E}}^n(f_H) \leq \underline{\mathbb{E}}^n(Rf_S) \leq \mathbb{E}^n(f_H) \leq \overline{\mathbb{E}}^n(Rf_S) \leq \overline{\mathbb{E}}^n(f_H); \text{ and}$$

$$\underline{\mathbb{T}}^n(f_H) \leq \underline{\mathbb{T}}^n(Rf_S) \leq \mathbb{T}^n(f_H) \leq \overline{\mathbb{T}}^n(Rf_S) \leq \overline{\mathbb{T}}^n(f_H).$$

Theorem 6.3 together will Lemma 6.1 implies that the escape of f_H through H is better approximated by considering any of its delayed first return maps Rf_S than by considering f_H itself. We now give a proof of Theorem 6.3.

Proof. For $S \in st(\Gamma_H)$, suppose $i \in M \setminus \mathscr{I}$ and $j \in \mathscr{I}$. Then

$$(\underline{A}_S)_{ij}(\mathbf{e}_S)_j = \begin{cases} \inf_{x \in \xi_{ij}} |f'(x)|^{-1} \mu(\xi_j) & \text{if } \xi_{ij} \neq \emptyset, \\ 0 & \text{otherwise} \end{cases} \leq \mu\{x \in \xi_i : f_H(x) \in \xi_j\}.$$

To show that a similar formula holds for larger powers of $\underline{A}_S$, suppose $k \in \mathscr{I}_S$. If $ik, kj \in \Omega_S$, then

$$(\underline{A}_S)_{ik}(\underline{A}_S)_{kj}(\mathbf{e}_S)_j = \inf_{x \in \xi_{ik}} |f'(x)|^{-1} \inf_{x \in \xi_{kj}} |f'(x)|^{-1} \mu(\xi_j)$$

$$\leq \mu\{x \in \xi_i : f_H(x) \in \xi_k, f_H^2(x) \in \xi_j\}.$$

If either $ik \notin \Omega_S$ or $kj \notin \Omega_S$, then $(\underline{A}_S)_{ik}(\underline{A}_S)_{kj}(\mathbf{e}_S)_j = 0$.

Suppose $k \in M \setminus \mathscr{I}_S$. If $ikj \in \Omega_S$, then $ikj; 1 \in M_S$ and

$$(\underline{A}_S)_{i,ikj;1}(\underline{A}_S)_{ikj;1,j}(\mathbf{e}_S)_j = 1 \cdot \inf_{x \in \xi_{ikj}} |(f^2(x))'|^{-1} \mu(\xi_j)$$

$$\leq \mu\{x \in \xi_i : f_H(x) \in \xi_k, f_H^2(x) \in \xi_j\}.$$

If $ikj \notin \Omega_S$, then $ijk; 1 \notin M_S$. In this case,

$$(\underline{A}_S^2)_{ij}(\mathbf{e}_S)_j = \sum_{k \in M_S} (\underline{A}_S)_{ik}(\underline{A}_S)_{kj} \mu(\xi_j)$$

$$= \sum_{ik,kj \in \Omega_S} (\underline{A}_S)_{ik}(\underline{A}_S)_{kj} \mu(\xi_j) + \sum_{ikj \in \Omega_S} (\underline{A}_S)_{i,ikj;1}(\underline{A}_S)_{ikj;1,j} \mu(\xi_j)$$

6.4 Improved Escape Estimates

$$\leq \sum_{k \in \mathscr{I}_S \cup (M \setminus \mathscr{I}_S)} \mu\{x \in \xi_i : f_H(x) \in \xi_k, f_H^2(x) \in \xi_j\}$$

$$= \mu\{x \in \xi_i : f_H^2(x) \in \xi_j\}.$$

Continuing in this manner, it follows that

$$(\underline{A}_S^n)_{ij}(\mathbf{e}_S)_j \leq \mu\{x \in \xi_i : f_H^n(x) \in \xi_j\} \qquad (6.16)$$

for $i \in M \setminus \mathscr{I}$, $j \in \mathscr{I}$, and $n \geq 1$. Since $(\mathbf{e}_S)_j = 0$ if $j \notin \mathscr{I}$, then for $n \geq 1$, equation (6.16) implies

$$\mathbf{1}_S \underline{A}_S^n \mathbf{e}_S = \sum_{i \in M} \sum_{j \in M_S} (\underline{A}_S^n)_{ij}(\mathbf{e}_S)_j \leq \sum_{i \in M \setminus \mathscr{I}} \mu\{x \in \xi_i : f_H^n(x) \in H\}$$

$$= \mu\{x \in I \setminus H : f_H^n(x) \in H\}.$$

Since $\mathbf{1}_S \underline{A}_S^0 \mathbf{e}_S = \mu(H)$, then $\underline{\mathbb{E}}^n(Rf_S) \leq \mathbb{E}^n(f_H) \leq \overline{\mathbb{E}}^n(Rf_S)$ for $n \geq 0$, where the second inequality follows by the same argument with the matrix $\overline{A}_S$.

To show that $\underline{\mathbb{E}}^n(f_H) \leq \underline{\mathbb{E}}^n(Rf_S)$, we again suppose that $i \in M \setminus \mathscr{I}$ and $j \in \mathscr{I}$. In this case, we have

$$(\underline{A}_H)_{ij}(\mathbf{e}_H)_j = \begin{cases} \inf_{x \in \xi_{ij}} |f'(x)|^{-1} \mu(\xi_j) & \text{if } \xi_{ij} \neq \emptyset, \\ 0 & \text{otherwise} \end{cases} = (\underline{A}_S)_{ij}(\mathbf{e}_S)_j.$$

For larger matrix powers, we have

$$\sum_{k \in M \setminus \mathscr{I}_S} (\underline{A}_H)_{ik}(\underline{A}_H)_{kj} \mu(\xi_j) = \sum_{k \in M \setminus \mathscr{I}_S} \inf_{x \in \xi_{ik}} |f'(x)|^{-1} \inf_{x \in \xi_{kj}} |f'(x)|^{-1} \mu(\xi_j)$$

$$\leq \sum_{k \in M \setminus \mathscr{I}_S} 1 \cdot \inf_{x \in \xi_{ikj}} |(f^2(x))'|^{-1} \mu(\xi_j)$$

$$= \sum_{ikj \in \Omega_S} (\underline{A}_S)_{i,ikj;1}(\underline{A}_S)_{ikj;1,j}(\mathbf{e}_S)_j.$$

From this, it follows that

$$(\underline{A}_H^2)_{ij}(\mathbf{e}_H)_j = \sum_{k \in \mathscr{I}_S} (\underline{A}_H)_{ik}(\underline{A}_H)_{kj} \mu(\xi_j) + \sum_{k \in M \setminus \mathscr{I}_S} (\underline{A}_H)_{ik}(\underline{A}_H)_{kj} \mu(\xi_j)$$

$$\leq \sum_{ik,kj \in \Omega_S} (\underline{A}_S)_{ik}(\underline{A}_S)_{kj} \mu(\xi_j) + \sum_{ikj \in \Omega_S} (\underline{A}_H)_{i,ikj;1}(\underline{A}_H)_{ikj;1,j} \mu(\xi_j)$$

$$= (\underline{A}_S^2)_{ij}(\mathbf{e}_S)_j.$$

Fig. 6.7 The delayed first return map $Rg_S : I \to I$ in Example 6.5, where Rg_S is delayed on the set ξ_{42}, shown in *red*

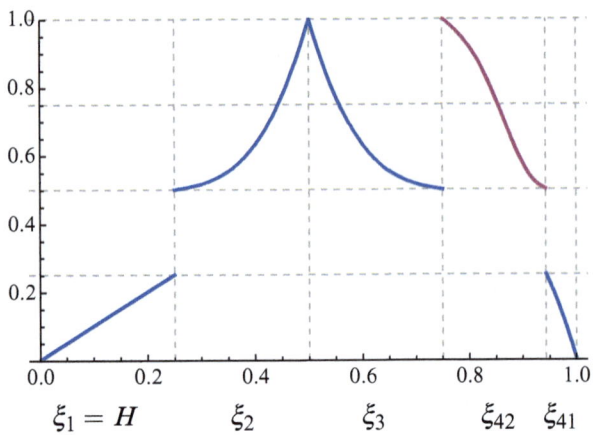

Again continuing in this manner, we have $(\underline{A}_H^n)_{ij}(\mathbf{e}_H)_j \leq (\underline{A}_S^n)_{ij}(\mathbf{e}_S)_j$ for $i \in M \setminus \mathscr{I}$, $j \in \mathscr{I}$, and $n \geq 1$. Since $\mathbf{1}_H \underline{A}_H^0 \mathbf{e}_H = \mu H = \mathbf{1}_S \underline{A}_S^0 \mathbf{e}_S$, it follows that

$$\mathbf{1}_H \underline{A}_H^n \mathbf{e}_H = \sum_{i \in M} \sum_{j \in \mathscr{I}} (\underline{A}_H^n)_{ij}(\mathbf{e}_H)_j \leq \sum_{i \in M} \sum_{j \in M_S} (\underline{A}_S^n)_{ij}(\mathbf{e}_S)_j = \mathbf{1}_S \underline{A}_S^n \mathbf{e}_S$$

for $n \geq 0$. Hence, $\underline{\mathbb{E}}^n(f_H) \leq \underline{\mathbb{E}}^n(Rf_S)$.

By using the same argument with the matrix $\overline{A}_S$, we obtain the inequality $\overline{\mathbb{E}}^n(Rf_S) \leq \overline{\mathbb{E}}^n(f_H)$. The second set of inequalities in Theorem 6.3 then follows, which completes the proof. □

Example 6.5. Consider the open system $g_H : I \to I$ given in Example 6.4. Observe that the vertex set $S = \{v_1, v_3, v_4\}$ is an open structural set of Γ_H, since v_1 is the vertex that corresponds to H, and the graph $\Gamma_H|_{\bar{S}} = \{v_2\}$ has no cycles (see Fig. 6.3).

The delayed first return map Rg_S is then written as

$$Rg_S(x_k) = \begin{cases} g_H(x_k) & \text{if } x_k \notin \xi_{42}, \\ g_H^2(x_k) & \text{if } x_k, x_{k-1} \in \xi_{42}, \\ x_k & \text{otherwise.} \end{cases}$$

The map Rg_S is shown in Fig. 6.7 as a one-dimensional map but is colored red on ξ_{42} to indicate that in fact, the system is delayed on this set. That is, $g_H(x)$ is shown in blue and $g_H^2(x)$ is shown in red, where the trajectories of Rg_S stay in ξ_{42} for two time steps before leaving.

6.4 Improved Escape Estimates

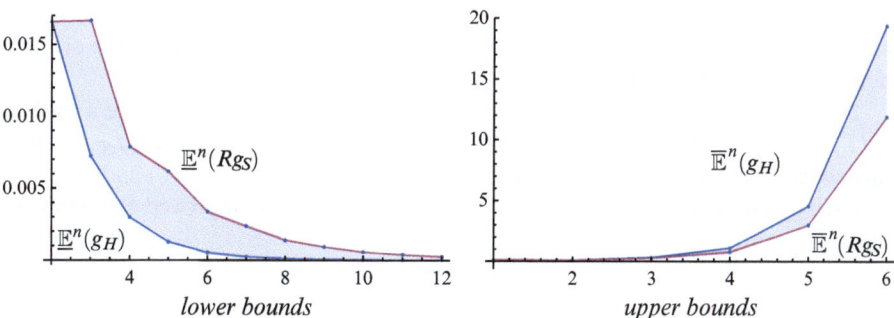

Fig. 6.8 Comparison between $\underline{\mathbb{E}}^n(g_H)$ and $\underline{\mathbb{E}}^n(Rg_S)$ (*left*) and $\overline{\mathbb{E}}^n(g_H)$ and $\overline{\mathbb{E}}^n(Rg_S)$ (*right*) from Examples 6.2 and 6.4

To compute the upper and lower transition matrices of Rg_S, note that the system's admissible sequences are given by $\Omega_S = \{23, 24, 33, 34, 424, 423, 41\}$, implying

$$M_S = \{1, 2, 3, 4, 424; 1, 423; 1\}. \tag{6.17}$$

From Ω_S, we compute that $\xi_{424} = (3/4, 0.85]$ and $\xi_{423} = (0.85, 0.94]$. The other partition elements have been computed in Example 6.4. Using the order given in (6.17), we have the upper and lower transition matrices

$$\underline{A}_S = \begin{bmatrix} 0 & 0 & 0 & 0 & 0 & 0 \\ 0 & 0 & 0.29 & 0.18 & 0 & 0 \\ 0 & 0 & 0.29 & 0.18 & 0 & 0 \\ 0.18 & 0 & 0 & 0 & 1 & 1 \\ 0 & 0 & 0 & 0.26 & 0 & 0 \\ 0 & 0.26 & 0 & 0 & 0 & 0 \end{bmatrix}, \quad \overline{A}_S = \begin{bmatrix} 0 & 0 & 0 & 0 & 0 & 0 \\ 0 & 0 & 4 & 0.29 & 0 & 0 \\ 0 & 0 & 4 & 0.29 & 0 & 0 \\ 0.29 & 0 & 0 & 0 & 1 & 1 \\ 0 & 0 & 0 & 0.73 & 0 & 0 \\ 0 & 1.18 & 0 & 0 & 0 & 0 \end{bmatrix}.$$

Here, the vectors $\mathbf{1}_S$ and $\mathbf{e}_S$ are given by

$$\mathbf{1}_S = [1, 1, 1, 1, 0, 0] \text{ and } \mathbf{e}_S = [1/4, 0, 0, 0, 0, 0]^T.$$

Figure 6.8 demonstrates that $\underline{\mathbb{E}}^n(g_H) < \underline{\mathbb{E}}^n(Rg_S)$ and $\overline{\mathbb{E}}^n(Rg_S) < \overline{\mathbb{E}}^n(g_H)$ for a number of values of n and indicates the extent to which using the delayed first return map Rg_H improves our estimates of $\mu(\mathbb{E}^n(g_H))$. The shaded regions in these graphs represent the differences between these respective upper and lower estimates.

Creating a delayed first return map Rf_S is similar in spirit to the procedure, described in Chap. 3, of constructing the time-delayed version $(\mathcal{H}_S\mathcal{F}, X_S^T)$ of the dynamical network $(\mathcal{F}, X)$. Recall that the time-delayed network $(\mathcal{H}_S\mathcal{F}, X_S^T)$ is created by absorbing the dynamics of those elements that are not indexed by $\mathcal{I}_S$, into those that are. The resulting time-delayed network is then restricted to the smaller set of elements S, which allows us to obtain better estimates of the original network's stability (see Definition 3.14 and Theorem 3.8).

Similarly, when we create the delayed first return map Rf_S, we choose a set $S \in st(\Gamma_H)$ over which we concentrate the dynamics of the open system (f_H, I). It is this concentration of information in the time-delayed system (Rf_S, I) and the time-delayed dynamical network $(\mathcal{H}_S\mathcal{F}, X_S^T)$ that allows for the improved estimates we obtain.

References

1. V. Afraimovich, L. Bunimovich, Dynamical networks: interplay of topology, interactions, and local dynamics. Nonlinearity **20**, 1761–1771 (2007)
2. V. Afraimovich, L. Bunimovich, Which hole is leaking the most: a topological approach to study open systems. Nonlinearity **23**, 643–656 (2010)
3. R. Albert, A.-L. Barabási, Statistical mechanics of complex networks. Rev. Mod. Phys. **74**, 47–97 (2002)
4. M. Blank, L.A. Bunimovich, Long range action in networks of chaotic elements. Nonlinearity **19**, 329–344 (2006)
5. R. Brualdi, Matrices, eigenvalues, and directed graphs. Lin. Mult. Alg. **11**, 143–165 (1982)
6. R. Brualdi, H. Ryser, *Combinatorial Matrix Theory* (Cambridge University Press, Melbourne, 1991)
7. L.A. Bunimovich, Ya.G. Sinai, Spacetime chaos in coupled map lattices. Nonlinearity **1**, 491–516 (1988)
8. L.A. Bunimovich, B.Z. Webb, Dynamical Networks, isospectral graph reductions, and improved estimates of matrices' spectra. Lin. Alg. Appl. **437**(7), 1429–1457 (2012)
9. L.A. Bunimovich, B.Z. Webb, Isospectral graph transformations, spectral equivalence, and global stability of dynamical networks. Nonlinearity **25**, 211–254 (2012)
10. L.A. Bunimovich, B.Z. Webb, Restrictions and stability of time-delayed dynamical networks. Nonlinearity **26**, 2131–2156 (2013)
11. L.A. Bunimovich, B.Z. Webb, Improved Estimates of Survival Probabilities via Isospectral Transformations, "Proceedings in Mathematics & Statistics" (eds. W. Bahsoun, C. Bose, G. Froyland) Springer, New York **70**, 119–135 (2014)
12. L.A. Bunimovich, A. Lambert, R. Lima, The emergence of coherent structures in coupled map lattices. J. Stat. Phys. **61**, 253–262 (1990)
13. J.-R. Chazottes, B. Fernandez, *Dynamics of Coupled Map Lattices and of Related Spatially Extended Systems*. Lecture Notes in Physics, vol. 671 (Springer, Berlin, 2005)
14. S. Chena, W. Zhaoa, Y. Xub, New criteria for globally exponential stability of delayed Cohen–Grossberg neural network. Math. Comput. Simul. **79**, 1527–1543 (2009)
15. F.R.K. Chung, *Spectral Graph Theory* (American Mathematical Society, Providence, 1997)
16. F. Chung, L. Lu, *Complex Graphs and Networks* (American Mathematical Society, Providence, 2006)
17. M. Cohen, S. Grossberg, Absolute stability and global pattern formation and parallel memory storage by competitive neural networks. IEEE Trans. Syst. Man Cybern. **SMC-13**, 815–821 (1983)

18. E.B. Curtis, D. Ingerman, J.A. Morrow, Circular planar graphs and resistor networks. Lin. Alg. Appl. **283**, 115–150 (1998)
19. S. Dorogovtsev, J. Mendes, *Evolution of Networks: From Biological Networks to the Internet and WWW* (Oxford University Press, Oxford, 2002)
20. M. Faloutsos, P. Faloutsos, C. Faloutsos, On power-law relationship of the internet topology ACMSIGCOMM,'99. Comput. Comm. Rev. **29**, 251–263 (1999)
21. S. Gershgorin, Über die Abgrenzung der Eigenwerte einer Matrix. Izv. Akad. Nauk SSSR Ser. Mat. **1**, 749–754 (1931)
22. R. Horn, C. Johnson, *Matrix Analysis* (Cambridge University Press, Cambridge, 1990)
23. M. Newman, A.-L. Barabási, D. Watts, *The Structure of Dynamic Networks* (Princeton University Press, Princeton, 2006)
24. S. Strogatz, *Sync: The Emerging Science of Spontaneous Order* (Hyperion, New York, 2003)
25. L. Tao, W. Ting, F. Shumin, Stability analysis on discrete-time Cohen–Grossberg neural networks with bounded distributed delay, in *Proceedings of the 30th Chinese Control Conference*, Yantai, China, 22–24 July 2011
26. L.N. Trefethen, M. Embree, *Spectra and Pseudospectra: The Behavior of Nonnormal Matrices and Operators* (Princeton University Press, Princeton, 2005)
27. R.S. Varga, *Gershgorin and His Circles* (Springer, Berlin, 2004)
28. F.G. Vasquez, and B.Z. Webb, Pseudospectra of isospectrally reduced matrices. Numer. Linear Algebra Appl. (2014). doi: 10.1002/nla.1943
29. L. Wang, Stability of Cohen–Grossberg neural networks with distributed delays. Appl. Math. Comput. **160**, 93–110 (2005)
30. D. Watts, *Small Worlds: The Dynamics of Networks Between Order and Randomness* (Princeton University Press, Princeton, 1999)
31. C.W. Wu, Synchronization in networks of nonlinear dynamical systems coupled via a direct graph. Nonlinearity **18**, 1057–1064 (2005)

Index

A
absolute row sum, 94
adjacency matrix, viii, xii, 1, 23, 73
adjacent cycle, 117
adjacent vertex, 30
admissible sequence, 164

B
basic structural set, 83
bounded radial transformation, 73, 78
branch, xi, 24
branch product, 24, 47
branch set, 24, 39, 45
Brauer-type region, 106
Brualdi-type region, 113

C
characteristic polynomial, 2
combinatorial Laplacian matrix, 123
complete structural set, 45, 76, 125
cycle, 23
cycle set, 112, 118

D
defective matrix, 154
degree of a rational function, 9
delayed first return map, 164, 166
directed graph, viii, 20, 21
dynamical network, ix, 53, 55
dynamical system, 53

E
edge set, 20
edge weight, ix, 20, 39, 46
eigenvalue multiplicity, 3
eigenvalue region, 91, 116
eigenvalues, viii, 2–4, 21
eigenvector, 130
escape, 147

F
final vertex set, 31, 34
fully reduced matrix, 102

G
Gershgorin-type region, 95
global attractor, 64
global stability, 55
globally attracting fixed point, 55
graph isomorphism, 36
graph of interactions, vii, ix, 20, 53, 72

H
hole, xiv, 147, 148

I
implicit time delay, 81
independent branches, 42
induced subgraph, 23
interaction, 55
interior vertices, 22, 24, 42

intrinsic stability, 59, 67, 68, 70, 71, 84
inverse eigenvalues, 3, 102, 139
inverse resolvent, 139
inverse spectrum, 3, 7, 15, 21, 102, 104
invertible matrix, 4
irreducible matrix, 73
isomorphic graphs, 36
isoradial expansion, 78
isoradial transformation, 73
isospectral expansion, 42, 46, 77
isospectral graph reduction, 19, 24, 28, 36
isospectral graph reductions, 31
isospectral matrix reduction, viii, 5–7, 10, 12, 91, 133, 136

L
local systems, 55, 62
loop, xi, 23
lower transition matrix, 165

M
Markov partition, 148
merged graph, 48
multiple-type delay, 54, 69, 70, 72

N
network expansion, 87, 89
network restriction, 83, 86, 87, 89
nondefective matrix, 154
nonnegative graph, 73
nonnegative matrix, 73
nonzero eigenvalues, 45, 49
nonzero spectrum, 45
normalized Laplacian matrix, 123

O
open dynamical system, xiv, 147, 148
open structural set, 164
orbit, 64
original Brualdi-type region, 115

P
parallel edges, vii, 48
path, 22
permutation index, 26
permutation matrix, 26
permutation matrx, 45
polynomial extension, 94
pseudoeigenvalue, 130, 144

pseudoeigenvector, 130
pseudoresonance, 139, 142
pseudospectrum, xiii, 129, 130, 132, 136

R
resolvent, 130, 132
resonance, 141

S
Schur complement, 5
semiring, 45
sequence of reductions, x, 12, 32–34
sequential reduction, 33, 34
simple graph, 123
single-type delay, 53, 69–72
spectral equivalence, xi, 36
spectral inverse, 15, 16, 102–104, 139
spectral radius, xii, 45, 49, 56, 60, 65, 73, 74, 125, 153
spectrum, viii, 1, 3, 7, 19, 21
stability, xii, 53, 55, 56, 61, 65, 71, 77, 84, 87, 89
stability matrix, 56, 59, 60, 65
stable, 68, 70
strong cycle, 112
strongly connected, 73, 112
strongly connected components, 73, 112
structural set, xi, 23
submatrix, 5, 8, 25
survival probability, xiv, 149, 159
symmetric matrix, 30, 124

T
time-delayed dynamical network, 64
time-delayed interaction, 64
tolerance, 130
topology, vii, ix, 20, 38, 53
transition graph, 150

U
undelayed dynamical network, 67
undirected graph, viii, 21, 30, 31, 123
unweighted adjacency matrix, 19
unweighted graph, 21, 24
upper transition matrix, 165

V
vertex set, 20, 23

W
weak cycle, 112
weight set, 21, 39, 45, 47
weight-preserving isospectral transformation, 39, 41, 42

weighted adjacency matrix, viii, 19, 21
weighted graph, 21
weighted transition matrix, 150, 159

MIX
Papier aus verantwortungsvollen Quellen
Paper from responsible sources
FSC® C105338

If you have any concerns about our products,
you can contact us on
ProductSafety@springernature.com

In case Publisher is established outside the EU,
the EU authorized representative is:
**Springer Nature Customer Service Center GmbH
Europaplatz 3, 69115 Heidelberg, Germany**

Printed by Libri Plureos GmbH
in Hamburg, Germany